Security Informatics and Law Enforcement

Series Editor

Babak Akhgar, CENTRIC (Centre of Excellence in Terrorism, Resilience, Intelligence and Organised Crime Research), Sheffield Hallam University, Sheffield, UK

The primary objective of this book series is to explore contemporary issues related to law enforcement agencies, security services and industries dealing with security related challenges (e.g., government organizations, financial sector insurance companies and internet service providers) from an engineering and computer science perspective. Each book in the series provides a handbook style practical guide to one of the following security challenges:

Cyber Crime - Focuses on new and evolving forms of crimes. Books describe the current status of cybercrime and cyber terrorism developments, security requirements and practices.

**Big Data Analytics, Situational Awareness and OSINT** - Provides unique insight for computer scientists as well as practitioners in security and policing domains on big data possibilities and challenges for the security domain, current and best practices as well as recommendations.

**Serious Games** – Provides an introduction into the use of serious games for training in the security domain, including advise for designers/ programmers, trainers and strategic decision makers.

Social Media in Crisis Management – explores how social media enables citizens to empower themselves during a crisis, from terrorism, public disorder, and natural disasters

**Law enforcement, Counterterrorism, and Anti-Trafficking** – Presents tools from those designing the computing and engineering techniques, architecture or policies related to applications confronting radicalisation, terrorism, and trafficking.

The books pertain to engineers working in law enforcement and researchers who are researching on capabilities of LEAs, though the series is truly multidisciplinary – each book will have hard core computer science, application of ICT in security and security/policing domain chapters. The books strike a balance between theory and practice.

Babak Akhgar · Petra Saskia Bayerl
Editors

# Accountability for Artificial Intelligence in Policing and Security

## From Concept to Practical Deployment

Springer

*Editors*
Babak Akhgar
CENTRIC (Centre of Excellence in Terrorism, Resilience, Intelligence and Organised Crime Research)
Sheffield Hallam University
Sheffield, UK

Petra Saskia Bayerl
CENTRIC (Centre of Excellence in Terrorism, Resilience, Intelligence and Organised Crime Research)
Sheffield Hallam University
Sheffield, UK

ISSN 2523-8507 ISSN 2523-8515 (electronic)
Security Informatics and Law Enforcement
ISBN 978-3-032-41283-6 ISBN 978-3-032-41284-3 (eBook)
https://doi.org/10.1007/978-3-032-41284-3

This work was supported by Sheffield Hallam University.



This Springer imprint is published by the registered company Springer Nature Switzerland AG
The registered company address is: Gewerbestrasse 11, 6330 Cham, Switzerland

*Dedicated to Aryan H. Akhgar. You are always in our hearts.*

# Foreword

Since its inception more than fifteen years ago, CENTRIC has been driven by a single, ambitious mission: 'to bring research into operational reality' for the benefit of policing and security. Throughout this journey, our work has consistently sought to bridge the gap between academic excellence and operational need, ensuring that research delivers tangible impact for practitioners, policymakers and society.

This book represents the culmination of several years of research and innovation on Artificial Intelligence, building upon a portfolio of research and development projects initiated in 2021. While AI continues to transform the policing landscape through enhanced analytical capabilities, automation, and decision support, its successful adoption depends not only on technological advancement but also on robust accountability, effective governance and public trust. These themes form the foundation of this volume.

The chapters provide an in-depth examination of accountability in the context of AI and policing and explore the opportunities and challenges presented by the increasing deployment of AI technologies across law enforcement. Rather than viewing accountability solely as a legal or ethical requirement or a conceptual instrument, this book argues that it should be recognised as a strategic capability that enables responsible innovation, strengthens organisational confidence, and reinforces public legitimacy.

A distinguishing feature of this volume is its strong emphasis on translating theory into practice. Drawing upon the outcomes of several national and international research programmes—including the Artificial Intelligence Policing Accountability System (AIPAS)—the contributors demonstrate how AI accountability can be operationalised through practical governance processes, structured assessment methodologies, and evidence-based implementation frameworks. The book illustrates these concepts through real-world examples spanning counter terrorism, organised crime, digital investigations, forensic science and wider policing applications.

Collectively, the chapters demonstrate that while conceptual models are essential for advancing academic understanding, responsible AI deployments in policing cannot rely on theory or thinking tools alone. AI accountability and AI governance must be embedded within operational processes, decision-making frameworks, organisational workflows and assurance mechanisms that reflect the realities of policing. This practical philosophy underpins the AIPAS guidelines and the broader body of work presented throughout this book.

The publication of this book is particularly timely as policing enters a new era of AI-enabled transformation. The UK Government's PoliceAI initiative represents a significant step towards the responsible, transparent and coordinated adoption of Artificial Intelligence across UK policing. Many of the principles explored in this volume—including governance by design, accountability by design, transparency, proportionality, explainability, human oversight and continuous assurance—closely align with the ambitions of the PoliceAI programme. In this respect, the research presented here not only supports the objectives of PoliceAI but provides practical methodologies, governance models and implementation guidance that can assist forces and national policing bodies everywhere in delivering AI safely, ethically and effectively.

Ultimately, this book is intended to serve both the academic community and operational practitioners. It offers researchers a rigorous examination of emerging concepts in AI accountability while providing police leaders, policymakers, technology providers and governance professionals with practical tools and frameworks that can be applied within real operational environments. We hope it will stimulate further discussion, encourage collaboration across disciplines and contribute to building an AI ecosystem in policing that is innovative, accountable, trusted and worthy of the public it serves.

As AI capabilities continue to evolve towards increasingly autonomous and agentic systems, the importance of robust accountability will only grow. Our aspiration is that this book contributes to shaping the next generation of accountable AI in policing—where technological innovation is matched by governance, transparency and accountability, ensuring that public trust remains at the heart of policing in the digital age.

Prof. Babak Akhgar, OBE, FBCS, FRSA
Director of CENTRIC

# Contents

PART I

# Foundations

CHAPTER 1

# Recognising the Benefits and Challenges of AI in Policing

*Christopher Rowley and Babak Akhgar*

## 1.1 Introduction

Artificial Intelligence (AI) is rapidly transforming the landscape of policing and internal security, emerging as a pivotal asset for enhancing the effectiveness and efficiency of policing and justice systems. Across a spectrum of roles and functions—from patrol officers and prosecutors to judges and border agents—AI allows the creation of innovative tools to carry out their purpose of upholding justice and ensuring public safety.

A significant number of police forces and agencies have begun to use AI in areas such as:

- **Facial Recognition**: Live retrospective AI-powered facial recognition technology is used at large-scale events such as concerts and sporting events to identify wanted offenders. Forces also have access to operator-initiated facial recognition, which is used by officers on

C. Rowley (✉) · B. Akhgar
CENTRIC, Sheffield Hallam University, Sheffield, UK
e-mail: christopher.rowley@shu.ac.uk



B. Akhgar and P. S. Bayerl (eds.), *Accountability for Artificial Intelligence in Policing and Security*, Security Informatics and Law Enforcement, https://doi.org/10.1007/978-3-032-41284-3_1

handheld devices to support them in confirming the correct identity of offenders.

- **Call Handling**: AI tools in call centres can analyse the sentiment in callers' voices to help operators prioritise responses based on urgency and vulnerability. This capability not only improves efficiency but also enhances overall satisfaction of callers by ensuring appropriate responses to individual needs.
- **Community Confidence and Trust**: AI provides real-time insights into public sentiment on social media, especially during critical incidents or major events. The early identification of emerging issues or rumours at the time can allow the police to deal with the issue promptly. This can demonstrate to the public that the police are aware and responding in a timely and appropriate manner.
- **Enabling Services**: Administrative policing functions such as finance, human resources, licencing, and procurement can use AI to streamline workflows, increase accuracy, and reduce costs. While this may lead to workforce reductions, it also presents opportunities to upskill employees, offering enhanced roles and remuneration. AI may also provide savings for police organisations, as and if it significantly reduces the number of IT systems required.

The drive to deploy AI in security and policing is a reaction to the growing usage of AI by criminal actors. Criminals and criminal organisations are becoming increasingly sophisticated, using secure communication channels to hide their online activity, obfuscating the movement of illicit funds, and leveraging AI to enhance their operations [26]. AI enables them to exploit new victims, and maximise their chances for profit in a shorter time. It is also being used to develop innovative criminal business models and reduce the risk of detection.

To counter these advancements, internal security and justice practitioners must adopt equally innovative tools, including AI, to safeguard society effectively [26]. Engaging in the use of AI may require new risk management strategies, as the potentials but also the risks of AI supersede previous policing capabilities. However, it must be recognised that failing to put adequate risk strategies in place also represents significant risks to public safety.

The groundbreaking potential of AI raises the critical challenge of employing AI's capabilities responsibly, capitalising on its benefits, while safeguarding accountability and protecting civil liberties. Members of the public and security practitioners alike express concerns regarding AI such as facial recognition and the potential for inbuilt bias. This may lead to the reinforcement of social inequality, the risk of faulty or unintended decisions and subsequent serious consequences, as well as the creation of rigid and opaque systems which overlook circumstances and in which decision-makers are unable to exercise discretion [30]. To maintain public trust, the use of AI must be regulated and guided by clear oversight to ensure algorithms are transparent, accountable and responsive to public and authority scrutiny. In fact, accountability is not just critical for trust but often a legal requirement for security and justice organisations [18].

This chapter will explain this dichotomy. It will explore the innovative prospects that AI offers in policing, while addressing the importance of clear guidelines and oversight mechanisms to ensure accountability, preventing misuse and protecting civil liberties.

## 1.2 AI Benefits to Policing and Law Enforcement

There are numerous potential benefits to be gained from AI in policing, including innovative investigative methods and its use in data analysis where AI models could be used to 'triage, prioritise, predict and manage data overload' ([17], p. 1). These benefits can extend across a variety of contexts, from routine policing activity through to critical crime areas.

Every day, police forces across the world engage in millions of interactions with the communities they serve, whether in person, over the phone, via email and text or through the recording of incidents using body-worn cameras. These interactions with the public generate vast amounts of data which, if recorded and managed properly, can be invaluable in preventing and detecting crime. Most police forces currently rely on intelligence analysts to sift through extensive datasets to create problem profiles, which assist in allocating resources effectively to prevent further offences taking place. AI has the potential to enhance this process significantly. With well-defined parameters, AI can analyse large data-sets to identify crime patterns, suggest potential suspects based on modus operandi and develop predictive patrol plans to place officers in the right locations at the right times.

In handling resource-heavy processes with speed and considerable precision, AI could not only reduce human workloads but also enhance operational accuracy and responsiveness. As a Scientific Adviser to the UK

National Police Chiefs' Council (local policing body) notes, AI integration allows officers to reduce time spent on bureaucratic functions and return to more traditional police activity [13]. These activities require human judgement and empathy, qualities essential to effective policing (College of Policing, 2024). Moreover, the use of AI in the judiciary could streamline evidence analysis, identify relevant case law and improve overall efficiency. In border security, AI tools already enable faster and more secure processing of passengers and transfer of essential data about their journey [19].

AI has also been highlighted as having future roles in supporting decision-making relating to children at risk of harm or with potential perpetrators of terrorism [17]. Beyond decision-making support, there are opportunities for AI technologies to be integrated into investigative functions and managing public communications. AI is progressively used in fingerprint analysis, increasing the speed of processing and identification (Chap. 12). Systems are increasingly capable of achieving accurate identifications from less-ideal samples, such as partial or smudged prints [22]. As this technology has progressed, there are indications that AI tools in future may be able to accurately identify biological sex through analysis of fingerprints [24].

Artificial Intelligence also promises critical developments in child protection. Research has long commented on police personnel involved in the identification and categorisation of Child Sexual Abuse Material (CSAM) being at increased risk of secondary trauma [4]. Developments in AI have resulted in multiple AI-driven tools to assist in the identification and categorisation of CSAM, with filtering technology speeding this process, automatically estimating age and minimising exposure through safe-viewing technologies [4]. These same technologies may also assist in the identification of child victims, through automatically grouping images of the same victim [21].

While such AI tools have immense potential in the fight against child abuse, AI has also been misused by offenders. Already in 2023, the Internet Watch Foundation (IWF) highlighted concerning activity on the dark web, with open-source AI image generation tools being trained to create CSAM [11]. Its 2026 report identified a further rise in risks from AI, with '8,029 AI-generated images and videos [...] assessed by IWF as showing realistic child sexual abuse' in 2025 [12]. This generation of AI-generated photorealistic CSAM has posed challenges to police forces [21]. Victim ID officers, seeking to focus their time on identifying real victims, now require methods to detect AI-generated material [11]. This may be complicated by a trend of offenders using AI to synthetically generate new CSAM of a particular real victim from existing catalogues, add the face of a real child, potentially one they know, onto existing CSAM, or 'nudify' non-indecent images of children [27]. This raises the prospect that ongoing development of criminal use of AI will require an innovative AI counter from policing and security agencies.

As these few examples show, as AI continues to develop, its potential to revolutionise policing and law enforcement grows exponentially. From saving lives and preventing crime to optimising resources during times of financial uncertainty, the opportunities AI presents are significant and will—if used ethically and appropriately—be transformative. Currently, the use of AI in policing is limited but expanding, and countries like the UK have put it at the heart of its policing reforms [20]. The UK Government perceives AI to represent a 'technological revolution' [29]. This allusion to the Industrial Revolution and its dramatic impact on society appears deliberate, with AI touted to 'transform nearly every aspect of our economy and society' ([29], para. 1).

In policing, this increasing development and deployment of AI represents a point of tension. The rapid ongoing advancement of technology requires a careful regulatory approach, needing consideration and, critically, time. Societal needs and concerns must be delicately balanced with the requirements of national security. Policing organisations thus face regular challenges that need to be addressed competently and proactively.

## 1.3 Common Challenges to AI in Policing and Law Enforcement

The integration of AI into policing must be understood within a broader socio-technical context, where technological systems interact with existing policing structures, legal frameworks and social inequalities. AI is not a neutral tool; rather, it reflects the data upon which it is trained and the priorities of those who design and deploy it [16]. Consequently, policing systems that rely on historical crime data may inadvertently reproduce or amplify existing biases, particularly those related to race, ethnicity, religion, political affiliation or geography. Predictive policing tools, for instance, have been criticised for disproportionately targeting some communities, thereby reinforcing cycles of surveillance and marginalisation [15]. The inappropriate application of predictive algorithms undermined public trust and confidence in police use of AI internationally [1].

This raises fundamental questions about fairness and legitimacy in policing. Public trust in law enforcement is closely tied to perceptions of procedural justice, which emphasises fairness, transparency, and impartiality in decision-making [28]. If AI systems are perceived as biased or unaccountable, they risk undermining this trust, even if they offer operational efficiencies. Therefore, ensuring that AI systems align with principles of procedural justice is essential for their long-term acceptance and effectiveness.

Another important consideration is the concept of 'algorithmic opacity', often referred to as the 'black box' problem. Many AI systems, particularly those based on machine learning, operate in ways that are not easily interpretable, even by their developers [3]. This lack of transparency poses significant challenges for accountability, especially in contexts where decisions have serious consequences, such as depriving someone of their liberty or direct surveillance. Without clear explanations of how decisions are made, it becomes difficult to challenge errors, assign responsibility and redress or ensure compliance with legal standards.

In addition, the effectiveness of AI is only as good as the quality of the data it processes. Poorly maintained or biased data can lead to flawed outcomes, perpetuating systemic issues or misidentifying individuals. Therefore, improving the accuracy and integrity of police data is a critical precursor to the widespread adoption of AI. Moreover, prior to the adoption of any AI system, the data used should be considered and fully cleansed, to ensure it is correct and free of errors and does not contain bias, which may cause the system to make recommendations that adversely affect or discriminate against particular groups of people or cause it to identify the wrong person for an offence.

Some police forces are already using robotic processes to clean their datasets, an important and necessary step which will ensure that AI outcomes (decisions and recommendations) are fair and reliable. In addition, each police organisation should ensure that all recommendations have human oversight, and subsequent actions should always be authorised by a human, who understands how the recommendations have been made, i.e. the source of the information and the implications of the possible actions that they will sanction. This will require additional training for decision-makers to ensure they understand how the AI system works.

There are also broader ethical concerns regarding surveillance and the potential for 'mission creep', where technologies introduced for specific purposes gradually expand into wider uses without adequate public debate or consent [25]. For example, systems initially deployed to identify serious offenders could later be used for lower-level offences or general population monitoring, raising questions about proportionality and necessity. This is a specific challenge, as many AI tools are in a 'permanent Beta state', and other legitimate uses will become apparent after their initial procurement. Ensuring that clear boundaries are established and maintained is therefore essential to prevent overreach and protect democratic freedoms.

Legal challenges have already begun to shape the trajectory of AI use in policing. A landmark case involving South Wales Police's use of live facial recognition found that aspects of its deployment were unlawful due to insufficient safeguards and a lack of clarity regarding data use and bias mitigation particularly with respect to the public sector equality duty [8]. The case of Bridges v South Wales Police [2] highlights the importance of embedding legal compliance and human rights considerations into the design and deployment of AI systems from the outset. It also demonstrates that public scrutiny and judicial oversight will play a crucial role in determining how far and how fast such technologies can be implemented.

Despite AI advancements, the deployment of AI in policing thus often remains uneven and experimental, with significant variation between forces in terms of capability, investment and governance structures. Some UK police forces, such as the Metropolitan Police Service and South Wales Police, have piloted or operationalised tools like live facial recognition, while others remain in early exploratory stages [27]. This fragmented adoption raises concerns about consistency in standards, oversight and public accountability across jurisdictions. Without a coordinated national framework, there is a risk that differing practices could lead to unequal impacts on communities and inconsistencies in the protection of civil liberties.

## 1.4 The Need for Oversight and a Policing-Specific Accountability Framework

In response to the challenges outlined above, there is increasing recognition of the need for AI accountability frameworks and governance models that can guide the responsible and accountable use of AI in policing.

Principles such as fairness, accountability, transparency and explainability (often referred to as 'FATE') are frequently cited as foundational to trustworthy AI [14]. Embedding these principles into procurement processes, operational policies and training programmes can help ensure that AI systems are not only effective but also aligned with societal values and legal obligations.

In consequence, public engagement is essential for ensuring that AI in policing aligns with societal values and expectations. This involves not only informing the public about how AI is used but also actively involving communities in discussions about its development and deployment (Chap. 10). Participatory approaches, such as public consultations or citizen juries, can help to identify concerns, build trust and ensure that diverse perspectives are considered [23]. Transparency in communication is particularly important in addressing misconceptions and fostering informed debate about the role of AI in policing.

In addition to transparency, robust governance frameworks are essential for ensuring accountability in AI-enabled policing. This includes the establishment of clear legal standards, ethical guidelines, and oversight mechanisms that define acceptable uses of AI and provide avenues for redress when harm occurs. In the United Kingdom, for example, bodies such as the Information Commissioner's Office (ICO) and the Centre for Data Ethics and Innovation (CDEI) play important roles in shaping the regulatory landscape for AI ([10]; see also Chap. 3). Internationally, frameworks such as the European Union's AI Act aim to categorise AI systems based on risk and impose stricter requirements on high-risk applications, including those used in law enforcement ([7]; see also Chap. 4).

Training and organisational culture also play a crucial role in the responsible use of AI. Police officers and practitioners must be equipped not only with technical skills but also with an understanding of the ethical and legal implications of AI tools. This includes recognising the limitations of these systems, avoiding over-reliance on automated outputs and maintaining critical judgement in decision-making processes. As scholars have noted, there is a risk of 'automation bias', where individuals place undue trust in automated systems, potentially leading to errors or unjust outcomes [9]. Mitigating this risk requires ongoing training, clear protocols and a culture that encourages questioning and accountability.

Ultimately, while AI offers significant opportunities to enhance policing capabilities, its successful integration will depend on the ability of institutions to address legal, ethical, and organisational requirements. A strong governance structure, which encourages a proactive transparent approach, supported by robust oversight and public engagement, will be essential in ensuring that AI contributes positively to policing outcomes without compromising fundamental rights.

## 1.5 Conclusion

Overall, the discussion in this chapter demonstrates that AI has many clearly defined benefits, which support the use of current AI systems and the internal development or purchase of new AI systems for policing. The AI systems are fast at analysing huge amounts of data in seconds and providing recommendations in the same amount of time, allowing officers and staff to act in a more effective and efficient manner; for instance, an early arrest using facial recognition, the prevention of further offending (i.e. predictive policing), or automating back-office functions to reduce costs. While each of these outcomes are welcomed and greatly needed by policing, each one is fully reliant on the quality of the data from which the AI system makes its decision, as well as the capability and understanding of the people that work with the AI system, and clear legal guidance of what is permissible.

There is therefore a compelling argument for the use of AI within policing and the wider law enforcement arena. However, there is also a great deal of uncertainty and fear about the reach of AI and its potential for providing incorrect outcomes. While it is accepted that there is a need for systems to deal with significant amounts of forensic data in complex investigations in a much timelier manner—and that there is a financial imperative to improve efficiency in enabling services such as Finance and HR—there is also the human element, which may slow down or even impede the adoption of AI, because individuals may fear the unknown or on a more personal level, fear they will lose their jobs to AI systems that can be more effective and efficient. There is also a need to ensure any AI systems that are developed in-house or purchased from third parties are both lawful and ethical when deployed, as each time a new system is used there is a significant amount of public scrutiny.

These are issues that Chief Officers will need to overcome, as there is an inevitability that policing will adopt AI through the combined necessities to reduce crime, improve outcomes for victims and reduce costs. As such the police organisation will need clear governance structures and monitoring in place to ensure the newly purchased or developed AI system is fit for purpose and is being used in a manner that complies with national as well as professional standards and the organisation's values. This level of oversight will assist the force in defending the use of AI, should there be an adverse reaction to an action which was derived from AI. It will also allow the organisation to identify where an AI system is adding value, where it can be improved and equally where it should not be used. Subsequent chapters in this book provide further discussions on AI accountability and practical recommendations in how elements of accountability can be accomplished, offering guidance to police leaders, police organisations and others in the practical implementation of AI accountability in policing.

**Acknowledgement** We gratefully acknowledge the support of Alexander Paradise for previous versions of this chapter. Work in this chapter was partially funded under the AIPAS project ('AI Accountability for Policing and Security'; UKRI Engineering and Physical Sciences Research Council, Responsible AI (RAI) IA091 Grant Ref: EP/Y009800/1).

## References

1. E. Bozkır, Y. Chen, S. Ghulam, Q. Ma, R. Razmi, R. Swaminathan, I. Thiru, Predictive policing or predictive prejudice? A study of the legal, historical and ethical implications of AI in policing. OxJournal (2025) https://www.oxjournal.org/predictive-policing-or-predictive-prejudice/
2. Bridges v South Wales Police, EWCA Civ, 1058 (2020)
3. J. Burrell, How the machine 'thinks': understanding opacity in machine learning algorithms. Big Data Soc. **3**(1) (2016). https://doi.org/10.1177/2053951715622512
4. C.M. Burns, J. Morley, R. Bradshaw, J. Domene, The emotional impact on and coping strategies employed by police teams investigating internet child exploitation. Traumatology **14**(2), 20–31 (2008)
5. College of Policinge, *Ethical Policing Principles* (2024)
6. Europol, *Internet Organised Crime Threat Assessment, 2021* (2021). https://www.europol.europa.eu/cms/sites/default/files/documents/internet_organised_crime_threat_assessment_iocta_2021.pdf
7. European Parliament & Council of the European Union, Regulation (EU) 2024/1689 of the European Parliament and of the Council of 13 June 2024 laying down harmonised rules on artificial intelligence and amending Regulations (EC) No 300/2008, (EU) No 167/2013, (EU) No 168/2013, (EU)

2018/858, (EU) 2018/1139 and (EU) 2019/2144 and Directives 2014/90/EU, (EU) 2016/797 and (EU) 2020/1828 (Artificial Intelligence Act). Official Journal of the European Union, L 2024/1689 (2024)
8. P. Fussey, D. Murray, *Independent Report on the London Metropolitan Police Service's Trial of Live Facial Recognition Technology* (2019)
9. K. Goddard, A. Roudsari, J.C. Wyatt, Automation bias: a systematic review. Med. Decis. Making **32**(1), 94–105 (2012)
10. Information Commissioner's Office, *Guidance on AI and Data Protection* (2020)
11. Internet Watch Foundation, *How AI is Being Abused to Create Child Sexual Abuse Imagery* (2023). https://www.iwf.org.uk
12. Internet Watch Foundation, AI-generated child sexual abuse: 2026 report on trends. Data Hum. Impact. (2026) https://admin.iwf.org.uk/media/hl1nvdti/iwf-ai-csam-report-2026.pdf
13. F. Knight, *Artificial Intelligence Already Working One Million Hours a Year for Police, as Concerns Raised Over Criminal 'AI Race'* (LBC, 2024, August 22). https://www.lbc.co.uk/news/artificial-intelligence-police-one-million-hours/
14. D. Leslie, *Understanding Artificial Intelligence Ethics and Safety* (The Alan Turing Institute, 2019)
15. K. Lum, W. Isaac, To predict and serve? Significance **13**(5), 14–19 (2016)
16. C. O'Neil, *Weapons of Math Destruction: How Big Data Increases Inequality and Threatens Democracy* (Crown, 2016)
17. M. Oswald, L. Chambers, A. Paul, Evaluating (semi)-autonomous systems in policing and national security: a new framework based on the concept of intelligence: a new matrix framework of evaluation and grading, based on lessons from existing processes designed to define and assess intelligence, in *Proceedings of the First International Symposium on Trustworthy Autonomous Systems.* (ACM, 2023), pp. 1–2.. https://doi.org/10.1145/3597512.3597524
18. Police Reform and Social Responsibility Act, 2011, c.13 (UK). https://www.legislation.gov.uk/ukpga/2011/13/contents/enacted
19. F. Sampson, *Convenience or Security—Which Came First?*. Biometric Update (2024, September 11) https://www.biometricupdate.com/202409/convenience-or-security-which-came-first
20. F. Sampson, *Official Recognition—'Transformative Technology' at the Heart of UK Police Reform.* Biometric Update (2026, February 1)https://www.biometricupdate.com/202602/official-recognition-transformative-technology-at-the-heart-of-uk-police-reform

21. L. Sanchez, C. Grajeda, I. Baggili, C. Hall, A practitioner survey exploring the value of forensic tools, AI, filtering, and safer presentation for investigating child sexual abuse material (CSAM). Digit. Invest. **29**, S124–S142 (2019)
22. R. Singh, R. Singh, R.K. Tripathi, P. Agarwal, Fingerprint recognition using artificial neural networks, in *Proceedings of the National Academy of Sciences, India Section A: Physical Sciences*, pp. 1–9 (2025)
23. G. Smith, *Democratic Innovations: Designing Institutions for Citizen Participation* (Cambridge University Press, 2020)
24. A.B. Spanier, D. Steiner, N. Sahalo, Y. Abecassis, D. Ziv, I. Hefetz, S. Kimchi, Enhancing fingerprint forensics: a comprehensive study of gender classification based on advanced data-centric AI approaches and multi-database analysis. Appl. Sci. **14(1), Article 417** (2024)
25. L. Taylor, What is data justice? The case for connecting digital rights and freedoms globally. Big Data Soc. **4**(2) (2017)
26. Trend Micro Research, *Malicious Uses and Abuses of Artificial Intelligence* (Europol, 2020) https://www.europol.europa.eu/cms/sites/default/files/documents/malicious_uses_and_abuses_of_artificial_intelligence_europol.pdf
27. A. Trivison, Understanding the line between art and abuse: How generative AI changes the landscape of child sexual abuse materials. Catholic Univ. J. Law Technol. **33**(1), 87–114 (2024)
28. T.R. Tyler, *Why People Obey the Law* (Princeton University Press, 2006)
29. UK Government, *Capabilities and Risks from Frontier AI: A Discussion Paper on the Need for Further Research into AI Risk* (2024). https://www.gov.uk/government/publications/frontier-ai-capabilities-and-risks-discussion-paper
30. M. Vivian, R.N. Ashrafur, M.T. Tusher, M.N. Akther, R.U. Rayhan, Ethical implications of AI-powered predictive policing: balancing public safety with privacy concerns. Innovatech. Eng. J. **2**(1), 47–58 (2025)


**Christopher Rowley KPM, MSt (Cantab)** retired as Lancashire Police's Chief Constable following a 31-year career in policing. A graduate of Cambridge University, he is passionate about identifying new and innovative ways of preventing people becoming victims by adopting evidence based academic evidence. He currently is Strategic Lead for Policing in CENTRIC (Centre of Excellence in Terrorism, Resilience, Intelligence and Organised Crime Research).

**Babak Akhgar OBE** is Professor of Informatics, Director of CENTRIC and a Fellow of the British Computer Society. He has extensive and hands-on experience in the development, management and execution of KM projects and large international security initiatives (e.g. combatting terrorism and organised crime, AI in Policing, cyber security, public order and cross-cultural ideology polarisation). He has more than 100 refereed publications in international journals and conferences on information systems, specifically focusing on knowledge management (KM). Babak was awarded an OBE in June 2022, the second highest order of the British Empire award, in recognition of his services to security research.

CHAPTER 2

# Police Accountability in Practice: The Example of UK Policing

*Christopher Rowley*

## 2.1 Introduction

A force that can demonstrate that it is both transparent in its dealing with the public and openly accountable to them will improve its prospect of winning their trust and cooperation. Public trust and support will be needed, as forces begin to fully embrace the use of AI. As detailed in other chapters in this book (Chaps. 1, 12, 13, 14), AI products can—and very probably will—revolutionise policing in the coming years. Yet if they are to do so, there needs to be a robust accountability process in place for the purchase (or development) of any new product, the way it is deployed and how it is used by operatives. This makes both the internal and external accountability process particularly important to ensure the necessary safeguards are in place to give the public, the media, the judiciary and politicians the confidence they need to support policing in its adoption of AI products.

C. Rowley (✉)
CENTRIC, Sheffield Hallam University, Sheffield, UK
e-mail: christopher.rowley@shu.ac.uk



B. Akhgar and P. S. Bayerl (eds.), *Accountability for Artificial Intelligence in Policing and Security*, Security Informatics and Law Enforcement, https://doi.org/10.1007/978-3-032-41284-3_2

This chapter will explain how statutory bodies, other organisations and the public hold the police to account in the UK. By outlining existing structures, processes and mechanisms that establish accountability, it provides an important view into the complex landscape of police accountability, into which AI deployments need to be integrated. Whilst the chapter considers accountability from a UK policing perspective, the principles are transferable to other policing bodies across the world.

## 2.2 Accountability in Policing

The integration of AI into policing and security raises profound questions about responsibility, oversight and public trust. To address these concerns, it is essential to establish frameworks that ensure accountability at every stage of AI deployment. Accountability in policing is a golden thread that runs from its inception by Sir Robert Peel to the current day. Whilst Peel did not explicitly use the term 'accountability', his principles, which articulate the role of the police and how they build and retain public trust and confidence, are clearly processes by which the police should hold themselves accountable—and conversely are the standards by which society will hold the police to account [2]. The Peelian principles were developed in 1829 in a very different era; however, they are still firmly embedded in the UK police culture. They are used and quoted on a regular basis by academics, politicians, the media and by policing organisations throughout the United Kingdom and beyond, when discussing policing, its role, and how that role should be delivered. It could be argued that the principles were the foundations on which police accountability has been derived.

The core function of the police is protecting life and property, preserving order and the King's peace in the UK, preventing the commission of offences and bringing offenders to justice. Most police funding for England and Wales comes from either central government grants or a local tax precept levied by the elected local policing body (see Sect. 2.3 below) and is paid by occupants in properties within defined policing areas. It is therefore arguable that the communities, which are served by the police, should have a mechanism to hold their local police force to account, particularly as they have no other option but to choose the one police force that covers their area.

The processes by which UK policing is held to account are numerous, including formal and direct forms of accountability. They affect local

Home Office forces (Police and Crime Commissioner) and the Mayor in other areas (e.g. London, the West Midlands, West Yorkshire, Greater Manchester and South Yorkshire). In addition to the Home Office forces, there are also other police forces, which provide most if not all of the services provided by a Home Office force, but they have additional requirements and different governance structures. They are the City Corporation for the City of London police (national fraud lead), the British Transport police authority for the British Transport police (responsible for policing the UK railway infrastructure), the Civil Nuclear Constabulary authority for the Civil Nuclear police (protection of the nuclear infrastructure across the UK) and the Ministry of Defence Police (responsible for protecting the UK national defence and related infrastructure), which is governed by the Ministry of Defence police committee on behalf of the Secretary of State. In Scotland, Police Scotland is governed by the Scottish Police Authority; members are appointed by Scottish Ministers, who are elected by the Scottish Parliament. In Northern Ireland, the Northern Ireland Policing Board governs the Police Service of Northern Ireland, members include representatives from the Northern Ireland Assembly, which is also elected. There are also police forces such as Teeside Harbour police and Mersey Tunnel police which have a very specific remit relating to a specific area. The following sections describe the current structures and mechanisms for both formal and informal forms of accountability in the UK.

### 2.2.1 *Mechanism of Formal Accountability*

Formal accountability refers to requirements and processes as set out in UK laws, regulations and policies on UK policing. By way of example, the primary role of the elected local policing bodies and Mayors in respect of Home Office forces in England and Wales is to hold the chief officer to account. Some of the powers conferred to the bodies under the Police Reform & Social Responsibility Act 2011, include hiring and firing the chief officer, ensuring the force is run in an effective and efficient manner that it delivers value for money and develops a police and crime plan. Whilst locally elected officials are the biggest actors in holding the chief officer to account, the Home Secretary is ultimately responsible for delivering policing across England and Wales. The relationship between the Home Secretary, the elected local policing body, Mayor and the chief officer is governed by the *Policing Protocol*, which is a statutory instrument that was laid before parliament (June 2023). The protocol is a helpful guide, which details how each office will work together and how they will deliver effective policing across England and Wales.

All local policing bodies and Mayors must develop and publish a *police and crime plan* or strategic plan, which will set out the objectives for the tenure of the elected individuals. However, the plan can be amended to respond to emerging crime or social issues. It is common practice prior to completing the plan that consultation will take place with key individuals, e.g. the chief officer and most importantly the communities they serve. This gives communities an opportunity to assist in developing the police and crime plan. The police and crime plan details the specific areas that the elected official requires the chief officer to deliver against. Examples of what is included in police and crime plans are issues such as anti-social behaviour, violence against women and girls, burglary and road traffic-related issues. Consulting communities and gaining their support in developing the contents of the police and crime plan is an extremely important part in its construction. By gaining that support, the elected official has an additional mandate to hold the chief officer to account as the voice of the people.

### 2.2.2 *Mechanisms for Direct Accountability*

Direct accountability refers to processes involving the public. *Community volunteers* can take the opportunity to work with local community policing teams to ensure they deliver against local needs, such as speeding or anti-social behaviour. This level of engagement is a daily occurrence in most police forces across the country. It is a form of direct accountability, in that the individual working with officers cannot only question their actions, but they can also give guidance regarding how they are exercising their powers within different communities. These interactions are often streamed on social media, playing an important role in demonstrating the actions of officers and how they are dealing with local community problems. Social media also allows for a two-way dialogue, giving the public the opportunity to hold officers to account for their actions or lack thereof.

Another effective way in which the public can hold policing to account is through the *complaints system*. This allows any member of the public to make complaints to their local police force, which under the Policing and Crime Act 2017 can be reviewed by the elected local policing bodies and Mayors who have their own statutory complaints system. The act also allows the elected bodies to have oversight of complaints. The purpose of these additional powers is to demonstrate further transparency to communities. Complaints can also be made to the Independent Office of Police Conduct (IOPC) or on social media or to the mainstream media. The gravity of each complaint is assessed and, where appropriate, a proportionate investigation is undertaken by the most appropriate body. Where the complaint is through social media or mainstream media, in most cases it will also be reviewed and assessed. The outcomes of the investigations or assessment will vary from national learning for all forces to individuals being personally held to account. The outcomes of all complaints will be constantly reviewed to establish if there is an issue with an individual, team, or if it is a force-wide problem. Where patterns emerge or the incident is very serious, a review will be conducted of the force's accountability processes to establish how this happened and how it can be stopped from re-occurring again.

### *2.2.3 Oversight Procedures*

To maintain public trust and confidence in policing and to ensure appropriate changes are made by local police forces, elected officials have very clear accountability processes which are often held in public. These are called *accountability meetings*. During these meetings detailed discussions take place regarding force performance against the objectives set out in the Police and Crime Plan, public complaints, force finance, partnership programmes, significant policing operations (where there are sensitivities relating to an operation, these discussions will often take place during the closed section of the accountability meeting), and any other issue relating to the effective, efficient and legitimate running of the organisation.

Whilst the elected official holds the chief officer to account at these meetings, often other members of the senior police team will be present and provide information relating to the questions posed. The frequency of accountability meetings is decided locally but often take place either monthly or quarterly. This meeting does not prevent other meetings taking place on a more regular basis to discuss issues at a more granular level. In England and Wales, the elected local policing body is also scrutinised on behalf of the community by a *police and crime panel*, which is made up of locally elected councillors. This is not needed where there is an authority in place. The elected local policing body reports on matters relating to the police and crime plan and finances. To assist the elected local policing body and their team, the chief officer can be invited and questioned by the panel, but they cannot hold the chief officer to account.

The UK Home secretary and the UK Home Office play an important part in holding the police to account in England and Wales. In most cases they provide the majority of funding for policing through the main grant for general policing and separate grants, which are specific to a particular initiative such as reducing violence against women and girls. The Home Secretary sets national and strategic direction and looks to ensure all the component parts of the policing systems work together to make communities safer. This is supported by the *Police Reform and Social Responsibility Act 2011*, which places an obligation on locally elected policing bodies to consider the strategic policing requirements (national threats, e.g. terrorism, violence against women and girls) when completing their police and crime plan. This process ensures local issues are aligned to national security and public safety requirements. They also hold chief officers and elected local policing bodies to account through the ***national policing governance and oversight boards***. During these meetings the Home Secretary will discuss performance across policing, they will also discuss forces who are national outliers in particular crime types and look for reassurance/plans, which will stop a further escalation of the problem to prevent further victimisation and to provide public reassurance.

The Home Secretary can hold police forces to account by utilising the resources of His Majesty's Inspectorate of Constabulary and Fire and Rescue (HMICFRS) to conduct detailed ***inspections of police forces***. The inspections can be thematic (e.g. firearms-related, super complaints such as the police response to stalking or the yearly Police Effectiveness, Efficiency and Legitimacy inspection—PEEL). When HMICFRS conducts a thematic inspection, for example, on reoffending, they may collaborate with other inspecting bodies such as HM Inspectorate of Prisons and HM Inspectorate of Probation. If the thematic is police-specific, HMICFRS's methodology is to select several forces and review the element they are inspecting across each force. This allows them to identify best practice or to identify themes which need addressing to improve service delivery.

After conducting the review, they will produce a report with recommendations; the individual forces will be named in the report, and recommendations will be made for all police forces across England and

Wales. *Super complaints* are complaints that are made by an organisation or group and are investigated by HMICFRS who will subsequently produce a report with recommendations. The PEEL inspection inspects the following nine areas at a granular level [1]:

1. Providing a service to victims of crime
2. Police powers and treating the public fairly
3. Preventing and deterring crime and anti-social behaviour and reducing vulnerability
4. Responding to the public
5. Investigating crime
6. Protecting vulnerable people
7. Managing offenders and suspects
8. Building, supporting and protecting the workforce
9. Leadership and force management

The scope of the inspection will dictate the length of time HMICFRS staff spend with the force. This can be several weeks for a thematic area to several months for a PEEL inspection. Most thematic reports are not graded; however, all PEEL reports are graded. Each of the areas detailed above (e.g. providing a service to victims of crime) will receive one of the following grades: 'Outstanding', 'Good', 'Adequate', 'Requires Improvement' and 'Inadequate'. All reports are published on the HMICFRS website and will be discussed with the elected official, e.g. the elected local policing body at the accountability board. *HMICFRS reports* are recognised as a key method of holding policing to account at a local and national level on behalf of the public and victims of crime. Such is the strength of the reports some have resulted in the dismissal of the chief officer.

## 2.3 Landscape of Police Accountability in the UK

### *2.3.1 External Accountability*

Whilst the public, elected officials and HMICFRS are the main actors in holding the police to account, Fig. 2.1 is a demonstration of other external organisations, who play a significant role in holding the police to account on behalf of the public.

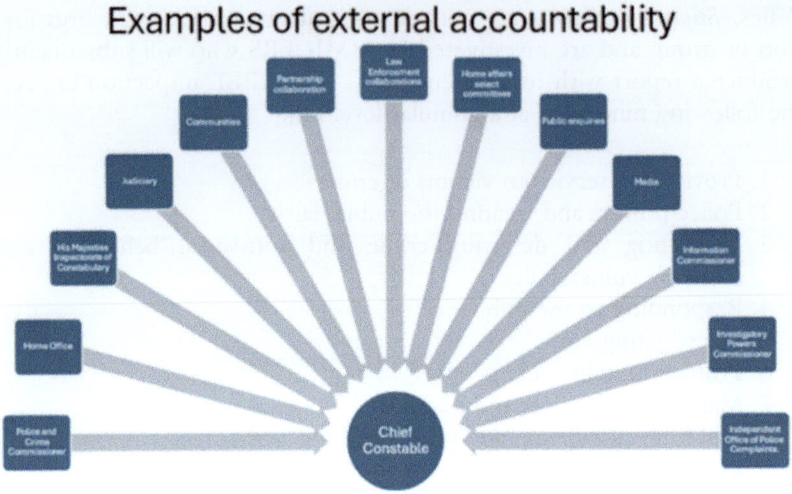


**Fig. 2.1** Bodies outside policing with roles in police accountability in the UK

All organisations detailed in Fig. 2.1, and many others, have different powers to hold the police to account but only the elected local policing body has an express statutory duty to do so. Their powers derive from acts of parliament such as the Information Commissioner and the Investigatory Powers Commissioner who conduct regular reviews. The *Information Commissioner* may conduct an investigation, where there has been a data breach (e.g. the release of personal information into the public domain). The findings are made public and often result in significant fines for the organisation. The *Investigatory Powers Commission* conduct yearly inspections of the use of intrusive and covert tactics by policing. The inspections consider, if forces are correctly authorising covert activity, ensuring any authority is legal, proportionate and necessary. It is the norm for both organisations to make detailed recommendations regarding their findings, which the force must act upon to prevent any further issues occurring.

Judicial findings in *courts*, be that criminal, civil suit, coronial, employment tribunal or an internal disciplinary panel are also forms of holding the force to account. Findings from hearings will cause the organisation to consider its processes and how those processes may have contributed to an event taking place, which may have had an adverse effect on an individual or the public more broadly. These findings will be considered and reviewed, if not through the elected officials, then by senior leaders in the organisation to prevent a reoccurrence and to learn any lessons. To ensure new processes and procedures are complied with and become the norm, accountability processes will be established to allow regular reviews to take place.

Social media and the mainstream media also play a significant role in holding policing to account daily. Social media is often used to capture police officers executing their powers where force is used. This will cause questions to be asked regarding the necessity and legitimacy of the force used against the individual. Events of this kind often result in policing conducting an internal enquiry or referring the organisation to the IOPC for an independent investigation, which when completed will generally be published. The mainstream media (i.e. the BBC, ITV and newspapers) often champion campaigns to highlight issues, such as violence against women and girls or knife crime in inner cities. These campaigns are often very effective, in that they not only highlight the problem, but they hold policing and government accountable for their actions in addressing the problem.

### *2.3.2 Internal Accountability*

Each of the areas and actors discussed above cause the police to constantly review their delivery models and their internal policies and procedures. The governance structure in any organisation is critical to delivering effective services, as is an open and transparent culture which regularly questions the quality of the service provided to the communities it serves. A critical element of a transparent culture is the internal accountability process, which underpins the governance of the organisation.

Most police forces have *Force Accountability meetings*, which are often chaired by the deputy chief officer. These meetings will be attended by the senior leaders within the organisation. The purpose of the meetings is to hold senior leaders to account for delivery against agreed plans, which are all linked to the police and crime plan, set by the elected local policing body after consulting with the public. At the start of the meeting, an overview of force performance will be presented; following this, a detailed discussion with senior leaders takes place to understand how performance can be improved. They also offer an opportunity for others to share best practice between Basic Command Units (BCU). Examples of discussions, which may take place are: how can a force reduce burglary offences, how can they answer and deploy to 999/101 calls quicker, and how can they reduce road traffic collisions? The discussions will concentrate on areas, which are outliers or causing significant harm to communities. At the conclusion of each meeting, which are normally a full day in duration, actions will be agreed and reviewed at the next force accountability meeting. Only actions agreed by the chair at this meeting or any other accountability meeting that have the requisite evidence to support their completion will be finalised.

The principles of the force accountability meeting are adopted at each level of the organisation. The assistant chief officer (who may be a civilian member of the organisation) will chair a *strategic meeting at a BCU* (operational delivery) and *departmental level* (e.g. dispatch centre, finance, professional standards). All aspects of local operational delivery will be reviewed to ensure all resources and finances are being used in an effective and efficient manner. Departments are also held to account in the same way, again to ensure they are providing the services required by the organisation to deliver an effective, efficient and legitimate police service. Accountability meetings also take place at a team level within each BCU and department. This allows senior leaders to review the productivity and compliance of their teams. An example of discussions, which will take place, are complaints by members of the public, sickness on the team, call handling times, detection of crimes and the health and wellbeing of the team. Supervisors also conduct individual accountability meetings. These meetings are designed to discuss the productivity of the individual, their individual development needs and any personal issues they may be dealing with.

*Accountability boards* will also be put in place to cover large change or delivery programmes within the force. Examples of these are a project board to develop new Target Operating Models, which are designed to remove waste, improve service delivery and often deliver multi-million-pound-saving targets. The local elected policing body is the legal owner of the police estate, and chief officers are unable to own real property. An Estates Board often oversees and delivers a strategic estates plan, which will ensure buildings are safe, in the correct area to service demand and where possible carbon neutral. The elected local policing body is also the contracting party in police procurement, even though management of the resources comes under the chief officer per the Policing Protocol Order. An IT board, amongst its strategic aims, will look to prevent successful cyberattacks and deliver an agile, effective and responsive IT system. Most forces will also have a HMICFRS accountability meeting to discuss any recommendations and progress against outstanding actions. All meetings will have a specific cycle (i.e. monthly, quarterly). The frequency will be dictated by the urgency of the recommendations or operational need. All meetings will be chaired, and all agreed actions will have an owner who will report on the progress of their actions at each meeting. Periodically, reports from each board will be presented to the chief officer team (COT), a meeting chaired by the chief officer. This allows the COT to review the progress of the board against agreed timelines and costings.

Whilst this internal accountability process may seem bureaucratic to some, it should be remembered that police forces have multi-million-pound budgets, they employ thousands of people, and answer, deploy and deal with a vast number of calls every day. They must ensure, as best they can, that their policies and procedures are considered, robust, effective, efficient and legitimate.

## 2.4 Relevance for AI in Policing

The above is an insight into both the external and internal influences, which either hold a force to account or cause a force with the correct governance to hold itself to account. AI in policing has relevance in two ways: firstly, as an area that requires accountability and oversight; secondly, as a capability that may support internal and external accountability procedures. For the latter, AI can provide the critical support needed to ensure good governance structures and are effective in large organisations, which are dealing with significant complexity on a daily basis. The correct AI systems will reduce the bureaucracy often associated with governance, by removing the administrative tasks linked to the monitoring and compliance of an effective and efficient governance structure. Through the correct application, AI can also highlight areas of both good practice and areas of concern within the structure which may affect the overall operational effectiveness of the organisation. For the former, the complex landscape of police accountability ensures that AI is already embedded in existing mature structures and processes. Still, throughout this book we argue that AI accountability is a specific thematic that needs targeted oversight and accountability considerations to be effective in practice (see Chap. 5).

The discussion in this chapter focused on the UK (primarily police forces in England and Wales) as an example of how police accountability is organised on local and national levels in a combination of formal and informal accountability, as well as internal and external structures. Whilst the exact procedures may be country-specific, the different forms of accountability will transfer to other countries and their AI deployment contexts. The UK example may therefore help in reviewing current accountability arrangements and support considerations how AI may affect these different accountability forms.

## References

1. His Majesty's Inspectorate of Constabulary and Fire & Rescue Services, *PEEL Assessments 2025–2027* (2025). https://hmicfrs.justiceinspectorates.gov.uk/peel-assessments/what-is-peel/peel-assessments-2025-2027/
2. Home Office, *Definition of Policing by Consent* (GOV.UK, 2012, December 10). https://www.gov.uk/government/publications/policing-by-consent/definition-of-policing-by-consent?utm_source=chatgpt.com

**Christopher Rowley KPM, MSt (Cantab)** retired as Lancashire Police's Chief Constable following a 31-year career in policing. A graduate of Cambridge University, he is passionate about identifying new and innovative ways of preventing people becoming victims by adopting evidence based academic evidence. He currently is Strategic Lead for Policing in CENTRIC (Centre of Excellence in Terrorism, Resilience, Intelligence and Organised Crime Research).

CHAPTER 3

# The Foundations of AI Accountability in Policing—UK Accountability Arrangements as Example

*Fraser Sampson and John McNeill*

## 3.1 Current AI Accountability Landscape

Extraordinary technological capability is coming to policing fast. If the citizen is going to trust its use for policing—and, in particular, for enforcing the law—artificial intelligence (AI) will need to bring comprehensive and consistent accountability with it. Police accountability is not always understood even by those involved in policing, and this chapter will set out some of the features by which accountability can be identified. This foundation will support the remaining chapters of this book, as they move through specific aspects, considerations and examples of AI systems used within policing, and flag the route to embedded accountability.

F. Sampson (✉) · J. McNeill
CENTRIC, Sheffield Hallam University, Sheffield, UK
e-mail: f.sampson@shu.ac.uk



B. Akhgar and P. S. Bayerl (eds.), *Accountability for Artificial Intelligence in Policing and Security*, Security Informatics and Law Enforcement, https://doi.org/10.1007/978-3-032-41284-3_3

> Accountability is about people with power using it properly and where those people are the police, their use of power can have profound consequences. If their power is amplified by technology, accountability extends to its use. In policing, advanced technology comes with advanced accountability. Sampson [11]

Police accountability is less about having the answer and more about having *to* answer (Chap. 2). The ways in which they have to answer, and the people to whom the police are accountable will vary from one country to another, but the general features will be very similar and are captured in the principles set out in the AIPAS framework (see Chap. 5). Police bodies must get ready to answer for their use of AI, and also for *not* using it in circumstances where it offers an available and legitimate tactical option in preventing and solving crime and keeping people safe.

Because AI capability is very different from that of previous policing technology, harnessing its potential will mean revisiting and re-engineering traditional answerability models.

## 3.2 Accountability Versus Legality

What is the difference between accountability and legality? That is a good question and a good place to start our journey through accountability in the use of Artificial Intelligence (or anything else) by the police. Accountability is closely connected with lawfulness, so closely in fact, that the other principles considered in this book (see Chap. 5) are really 'Lawfulness *Plus*'.

At a basic level, all aspects of the use of AI should be governed by formal, promulgated rules. It extends to all those involved in building, developing and operating AI systems for use in a criminal justice context. Where any gaps in the law exist, the protection and promotion of fundamental rights and freedoms should prevail. At the same time, policing bodies are emanations of the state, exercising exceptional powers and enjoying extraordinary privileges. Where they are interfering with the fundamental rights and freedoms of the citizen the state and its emanations must be able to point to legal authority to conduct operations. In this respect, being answerable to the law and the courts forms part of police accountability and there are clear accountability pathways in place, some differing between jurisdictions in the UK and others applying to all.

Interestingly, some policing bodies are called 'police services' (e.g. Scotland and Northern Ireland), while the statutory inspection body's title refers to the 'constabulary'. England and Wales have an arrangement of geographically aligned police 'forces'. Whatever the origin and impact of nomenclature, all policing bodies have clear legal obligations, rights and remedies governing the conduct of all public bodies, together with many private law considerations around the use of products and services in a way that causes loss or harm to individuals. All these legal considerations will form part of the accountability 'picture' for the police use of AI.

In the principal case, addressing some of the issues of police use of AI-enabled capability (live facial recognition) the Court of Appeal[1] highlighted the importance of applying accountability to surveillance technologies.

While some might argue that the police use of AI could rely on the existing legal tests—for example, in the use of intrusive surveillance technology and storage of data[2]—there will be situations where the technology has fundamentally altered the considerations. For example, AI-driven retrospective facial matching capability that is now available to all UK policing bodies elementally alters the ability to search, edit and store data from which individuals can be recognised in perpetuity. This makes the level of potential intrusiveness wholly and materially different from any technology that might have been considered by the courts previously.

Despite many public criticisms of the perceived inability of the law to keep pace with AI-driven technology,[3] the courts in England and Wales have made clear that 'the fact that a technology is new does not mean that it is outside the scope of existing regulation, or that it is always necessary to create a bespoke legal framework for it'.[4] However, while this means that new technologies may be able to integrate into the co-existing common law rules that govern policing and security, the extent

[1] R (on the application of Bridges) v Chief Constable of South Wales (2020 EWCA Civ 1058).

[2] R (on the Application of Catt) v Commissioner of Police of the Metropolis and Another UKSC/2013/0114.

[3] https://hls.harvard.edu/today/is-the-law-playing-catch-up-with-ai/.

[4] Bridges at 84.

of novelty in AI-powered technology may mean there is no precedent to apply—we have not been here before.

An interesting area for the development of the law in this area will be the tortious duties and remedies applying to products and services. While AI tools and their developers will attract the general duties applicable to commercial and individual uses,[5] the police in England and Wales owe no duty of care to the population at large,[6] and it is unclear how far autonomous systems acting as agents of the police might be interpreted in a way that disturbs that settled situation. The law may also develop in a different direction in other UK jurisdictions (see sect. 3.6).

Recent developments in relation to the impact of Art 3 of the European Convention on Human Rights (ECHR) illustrate how some of the first AI-related challenges to established law may come about. In *Commissioner of Police of the Metropolis v DSD and Another*,[7] the Supreme Court found the police liable for a failure to protect citizens from degrading or inhumane treatment at the hands of an offender whom they were investigating for serious sexual offences. The Supreme Court held that the police have a positive obligation under Art 3 to conduct an effective investigation of crimes resulting in this treatment of the citizen, even where the cause of that treatment is a private individual rather than the state. The basis for liability in *DSD* was partly the investigative 'means' available to the police by which they might have reasonably prevented the harm to subsequent victims of a serial rapist and sexual offender. There is clear room for this duty to be developed by the courts, as AI-enabled preventative technology such as facial recognition technology (FRT) becomes more readily available to the police and it is reasonable to expect that the police will one day have to provide a legal defence for their decisions *not* to use AI-enabled technology as often as they currently do for using it.

In terms of the technology and equipment used by the police there are several further considerations for meeting standards in discharging legal obligations. For example, the British Standards Institute has produced a core framework related documentation surrounding the usage of AI aimed at an array of components involved in the AI lifecycle. The British

[5] See e.g. Caparo Industries plc v Dickman [1990] UKHL 2.

[6] Hill v Chief Constable of West Yorkshire [1987] UKHL 12.

[7] UKSC 2015/11/0166.

Standard ISO 42001:2023 intends to provide a 'certifiable AI management system framework', which states that if an organisation was to conform the requirements of the BSI it can 'generate evidence of its responsibility and accountability regarding its role with respect to AI systems'. This situation, that the UK policies and procedures frameworks face, is a result of there being no enshrined legal requirement to form an evidence base. By contrast, the EU AI Act (see Chap. 4) is an example of a legislative policy to address the potential powers and risks of AI across international boundaries. Another area is compliance with the Forensic Science Regulator's statutory code of practice in England and Wales.[8] But embedded accountability goes much further than legal compliance.

## 3.3 Accountability in UK Policing

Accountability is measured in part as the prevention and redress of abuses of power, but it also aims to understand expectations of stakeholders that in-turn could lead to ethical or societal impacts [1]. Accountability is defined in AP4AI (Accountability Principles for Artificial Intelligence; [1]) as the extent to which practitioners are generally accountable to their communities, in other words it is a proxy measure for the extent of their *legitimacy* within those communities (see Chap. 4).

The overlap between law and accountability is clearest in the establishment of governance and oversight bodies for the police. As creatures of statute, police forces' accountability arrangements are themselves enshrined in legislation. The police in the United Kingdom have formal legal accountability structures linking them to their local communities. Of the 43 geographically defined police forces of England and Wales, 41 operate under a clear statutory framework for accountability to an elected individual officeholder. The Police Reform and Social Responsibility Act 2011 (PRSA)[9] sets out what the forces are and how their chief officers will be accountable. The PRSA requires every police force to have a directly elected individual to hold the chief officer to account and in many large areas of the country—including London—the elected local policing body is the Mayor; most will designate an elected deputy mayor for policing

[8] https://www.gov.uk/government/publications/statutory-code-of-practice-for-forensic-science-activities/forensic-science-regulator-code-of-practice-accessible#bookmark5.

[9] The City of London Police have a police authority structure operated by the Corporation.

and crime. In addition to financial responsibilities and for setting a local tax precept for police funding, the elected body will invariably be the contractor for procurement of equipment such as IT and vehicles and is the only policing body empowered to own land or an interest in real property. This accountability mechanism is intended to establish a direct democratic mandate for a locally elected official to hold their policing bodies to account on behalf of the citizen (for more details on policing accountability see also Chap. 2).

While sworn police officers enjoy a wide degree of operational independence from their elected oversight bodies, and the respective functions of chief officers and their oversight bodies are set out in specific legal instruments,[10] the chief officer's accountability to their elected local policing body extends to operational policing. This was made clear by the High Court which held that the elected policing body is not just entitled, but *obliged*, to hold the relevant chief officer to account in respect of all their functions and for all the functions of those acting under their direction and control.[11] Those functions will include AI-relevant activities from procurement to obtaining value for money and the way in which it is used.

## 3.4 Non-geographic Policing Bodies

There are several policing bodies with sworn constables that do not come under the above governance arrangements. Some are referred to as statutory undertakers because they have specific legal undertakings within their operational remit, such as licensed civil nuclear sites and the UK's rail network.

The operational activities of these bodies will integrate with those of the geographical policing bodies described above to varying degrees. Often it may be unreasonable to expect the citizen to detect, let alone understand, any legal or organisational differences when dealing with officers and staff from combined operations. The importance from an accountability point of view is to note that the fundamental elements of responsive and responsible governance that apply to UK policing generally will apply equally to these specialist police bodies. While their

[10] See the Policing Protocol Order, SI 2011/2744.

[11] R (on the Application of Crompton) [2017] EWHC 1349 Admin at 76.

legal constitution and oversight arrangements will be different, the below organisations form part of the police services for the UK and will generally be covered by national operational policies and professional practice such as that issued by the College of Policing[12]:

- British Transport Police
- Civil Nuclear Constabulary
- Ministry of Defence Police

Each of the above police services has a clear statutory governance framework in place and works very closely with other relevant policing bodies throughout the UK and abroad. In addition, the UK has a National Crime Agency (NCA)[13] which operates under a different accountability regime from the policing bodies with which it works closely. The Agency has significant international responsibilities and therefore must take close account of jurisdictional issues, particularly the application of legislation governing the use of AI (such as the EU AI Act). The issues, principles, and risks/mitigations discussed throughout this book apply equally to their activities and uses of AI for policing and law enforcement.

All the police services, whether geographic or statutory undertakers, will come within the oversight responsibilities for policing generally, including a specific inspectorate for policing, fire and recuse services (HMICFRS[14]) and the Independent Office for Police Conduct established under the Police Reform Act 2002,[15] while regulatory frameworks governing public bodies more widely such as Freedom of Information[16] and Subject Access Requests[17] also apply to the police. But the existence of legal obligations and avenues of influence from the citizen is not synonymous with comprehensive accountability.

12 https://www.college.police.uk/app.

13 https://www.nationalcrimeagency.gov.uk/.

14 https://hmicfrs.justiceinspectorates.gov.uk/.

15 https://www.policeconduct.gov.uk/.

16 See the Freedom of Information Act 2000.

17 Under the Data Protection Act 2018.

Here is an AI-focused example. The police use of FRT is a contentious and ongoing issue of considerable public concern [10]. Freedom of information requests from citizens to ask their local police for data about the use of facial recognition technology showed that the police did not collect that data and as a result the police refused to provide the information as the cost of collating it would be excessive (a statutory exemption). The responses also showed that the police did not consult with the UK's data regulator.[18] Aside from how such 'refusals' appear to the public when trying to exercise their statutory accountability powers, without this information the police are not able to assure *themselves* of the AI-enabled system's legitimate use. Moreover, the lawful, proportionate, and justified use of police biometrics is arguably far too important to be left to the voluntary exercise of statutory avenues by the individual citizen. In securing and retaining the trust and confidence of the public in new AI capabilities the police will need to be proactive, demonstrating that they are alive to the importance of accountability and making their case in a clearer, more evidence-based way (see also Chap. 10). Increasingly there will be cases where the public expect the police to use retrospective facial capabilities—identifying victims and offenders in child sexual abuse image cases for example—and the police will find themselves explaining decisions *not* to use the technology. That is accountability, perhaps a different form of accountability than they are used to, and it will need reliable and consistent performance data.

Beyond these arrangements for ensuring general police accountability, there are some broader AI-specific issues to be aware of.

## 3.5 AI Accountability in Policing

Over and above any organisational accountability arrangements, some commentators have argued that AI needs a 'clear accountability process and ecosystem' ([2], p.132) in place, while the UK government has recognised the need for 'clear lines of accountability across the AI life cycle' [4]. In this AI-specific context the accountability element it seen as both a tool to provide 'transparency and human oversight of both the actions taken by the agent and the inference performed by the system'

[18] https://www.computerweekly.com/news/366589095/Police-Scotland-did-not-consult-ICO-about-high-risk-cloud-system.

([8], p. 6) and also a method of highlighting individual accountability for misuse.

Some academics have argued that the UK government should be providing more methods to ensure greater transparency and accountability around the use of algorithmic systems in public sector decision-making [3]. In the peculiar setting of policing and law enforcement the accountability issues become even more focused [11].

Description has proved problematic for the government. For example, a Department for Science, Innovation and Technology (DSIT) document states that 'AI systems should not undermine the legal rights of individuals or organisations, discriminate unfairly against individuals, or create unfair market outcomes' (UK Government 'Introduction to AI Assurance' [4], p. 10). Axioms like this do not really advance the challenge of how to achieve accountability—take that same sentence and insert the name of any other police technology—Taser, DNA profiling, body worn video—in the sentence and it does not change. (Further discussion of the concepts for the purposes of accountability can be found in Chap. 4).

## 3.6 Effective AI Accountability in Policing

What is the best measure of police effectiveness? According to the founder of modern British policing, it is the absence of crime and disorder, not the visible evidence of police action in dealing with them—one of the so-called Peelian principles. While it is unlikely that Robert Peel came up with them himself [9], his nine principles have remained sacrosanct in many Western jurisdictions for over 150 years. But what is the best measure of police *accountability*?

Some of the Peelian principles do not fit within today's AI-enabled policing world. For example, his second principle says the police should demonstrate impartial service to law, *in complete independence of policy* when policy is a central part of the lawfulness of AI-enabled biometric systems such as FRT [*R (on the Application of Bridges) v Chief Constable of South Wales Police & Ors* [2020] EWCA Civ 1058; *R (Thompson and Carlo) v the Commissioner of Police of the Metropolis and the Equality and Human Rights Commission* [2026] EWHC 915 (Admin)].

Artificial intelligence presents a different, more demanding, accountability challenge for policing, a challenge partly determined by what the police are doing with it. If using elemental AI for purely administrative functions, the police have the same level of accountability as other state agencies. If a chief officer wants to use AI to issue uniforms that is no different from the local authority doing it, but using AI for *operational policing functions* is a wholly different use case. A chief officer using AI to order boots off the shelf is not the same as ordering boots onto the street. Where the technology is used for a law enforcement purpose like remote biometric identification surveillance policing must anticipate deeper levels of accountability. Extending that to inferential algorithms calculating age, mood and so on raises the stakes irrespective of who is using AI, and the operator will need to reflect the legal refinements (such as the EU AI Act which may render it unlawful) and reinforce any necessary safeguards.

With AI being in an almost perpetual *Beta* state, the established accountability such as vigilance over 'function creep' and purpose extension must also be revisited. It is likely that uses for an AI system will emerge that were unforeseen at the time it was purchased. Function or mission creep will be a challenge for accountability bodies who will have to strike a new and dynamic balance between legitimate diversification and unapproved experimentation.

As AI tools become more established and accepted in policing, it will be important not to conflate arguments that the AI technology is itself flawed or unlawful rather than the way in which the police use it. For example, one of the grounds on which the police use of live facial recognition technology in *Bridges* (*loc cit*) was found to have been unlawful was because of their failure to discharge their general, non-AI duties under the Equality Act 2010; the use of the technology was, of itself, not a breach of the duty to act 'in accordance with law'.

Academics have also argued that FR technology poses a threat to freedom of expression, assembly and association, and the right to privacy [7]. The principle of legality is core to providing safeguards to human rights as highlighted in other frameworks that argue the principle of legality (The European Commission for Democracy; [6]). This re-enforces the requirement that clear legal guidelines are a requirement for AI to mitigate the risk of misuse or negative impact against vital tools for individuals in the police.

It is not only clear but predictable that the law is 'notoriously slow-moving in comparison to the pace of change and direction of technology' [5]. The legislative process necessarily involves stages of careful scrutiny and amendment; yet a view that fast-evolving technology requires fast-evolving law persists. The answer may be for the legislative instruments regulating AI to keep up, but for an illustration of what a footrace between criminal legislation and science looks likes, take a look at the Misuse of Drugs Act 197. That legislation outlaws specified substances as 'controlled drugs' as set out in its schedules which, as new drugs are created, are amended. After amendment, the drug recipes are amended by chemists and a new variant created which is then added to the schedule. Take the infinite potential that AI offers, and it is easy to dispel the argument that the only solution is for a relative acceleration of legislation.

Aside from the sheer practical impossibility of such an approach, in the realms of digital technology, people generally expect software to be patched regularly but not the policy or legislation that govern it. Which is why a set of clear and flexible principles was felt to be more suited to regulating AI as set out in the AI accountability concept put forward in this book and formulated in the AI Accountability Guidelines for Policing and Security (AIPAS; see Appendix).

Effective accountability for AI within policing and law enforcement will ultimately depend on balancing what is technologically possible, what is legally permissible, and what is societally acceptable—in other words balancing innovation, regulation, and expectation. How will that be measured? A good start would be the application of AI accountability principles.

## 3.7 National Variations within the UK

Scottish law is a separate—although substantially overlapping—body of jurisprudence from that of England and Wales. Road traffic laws are very similar, as are many of the basic concepts of policing and constitutional law. In Northern Ireland many of the police powers and governance arrangements for the police are similarly worded to those in England and Wales. However, both jurisdictions have significant differences in their legal and wider governance oversight arrangements. Both Scotland and Northern Ireland have chosen to describe their national police as 'services' rather than 'forces', which is the term used in England and Wales.[19] Both have also established a tripartite arrangement between the operational policing service, the statutory authority with responsibility for the maintenance of an effective police service, and an emphatically human rights-focused complaints and misconduct framework. The third of these

[19] Cf. the Civil Nuclear 'Constabulary' above.

features allows fewer derogations from the applicable human rights provisions of the UK and is regarded as a strengthening factor for police accountability in comparison to England and Wales. With some urging a dilution of international human rights influence generally within the UK,[20] it will be interesting to see how any change in policy or legislation will impact on the policing services of these two jurisdictions.

For the purposes of this book a few key features of the accountability arrangements have been highlighted but this does not attempt to be anything more than a descriptive illustration.

### *3.7.1 Scotland*

While England and Wales have directly elected local policing bodies (see above), Scotland and Northern Ireland have different forms of democratic accountability. The statutory accountability arrangements for policing in Scotland are different from those in England and Wales, and there is no equivalent of the elected local policing bodies (mayors and police and crime commissioners).

The Scottish Police Authority (SPA) is established under the Police and Fire Reform (Scotland) Act 2012 and came into operation on 1 April 2013, along with the Police Service of Scotland. The SPA's members are appointed by Scottish ministers, who are elected by the Scottish Parliament.

The core functions of the SPA are defined in the Act which sets out a legislative framework to ensure effective governance for policing services across Scotland with the SPA being responsible for providing national governance and accountability arrangements.

In carrying out its functions the SPA has specific regard to the SPA's Governance and Accountability Framework Document drawn up by Scottish Government in consultation with the SPA. The framework document describes the context within which the SPA operates and sets out key roles and responsibilities. Other statutory elements of accountability and governance include the Public Finance and Accountability (Scotland) Act 2000 and the Ethical Standards in Public Life etc. (Scotland) Act 2000.

The SPA also operates under various regulations and directions issued by the Scottish Ministers, with additional obligation arising from data

[20] https://www.reuters.com/world/uk/britain-under-pressure-immigration-urges-reform-european-human-rights-framework-2025-06-18/.

protection and other UK-applicable legislation like the Data Protection Act 2018 and UK GDPR. While Scotland has its own independent scrutiny body,[21] and a commissioner for police use of biometrics,[22] matters like counter terrorism and national security are reserved and the relevant functions of UK bodies will apply.

The Police (Ethics, Conduct & Scrutiny) Scotland Act 2025 (PECSS) was enacted by the Scottish Parliament following the Angiolini Review of Police Complaints Handling, Investigations and Misconduct Issues. The PECSS covers a range of topics related to ethics and conduct for policing, as well as the functions and governance of the Police Investigations and Review Commissioner (PIRC) and supports independent scrutiny by PIRC and SPA. One significant difference in the accountability arrangements for professional conduct is the existence of a duty of candour for Scottish police officers, something that has been proposed for public bodies in the UK including the police for many years and which the UK government has undertaken to introduce during the current parliament.[23]

In relation to the civil liability of the police for using/failing to use AI systems, the basis for assessment may be different in Scotland from that of England and Wales. The existence of a duty of care owed by the police to the population as first established in *Hill* (above) may be approached differently by Scottish courts where the creation of such a duty can arise from police activity (such as failing to protect a dangerous stretch of road over which they have taken control following an incident[24]).

### *3.7.2 Northern Ireland*

Like Scotland, Northern Ireland has its own single police service. The legal arrangements underpinning the accountability of the police are more complex than those for the other parts of the UK. Again, this part of the chapter is simply a summary of some of the key elements.

[21] His Majesty's Inspectorate of Constabulary in Scotland.

[22] https://www.biometricscommissioner.scot.

[23] https://commonslibrary.parliament.uk/research-briefings/cdp-2025-0171/

[24] Gibson (AP) v Orr, Strathlclyde Police Chief Constable [1999] ScotCS 61 (26 February 1999).

The Police Service of Northern Ireland (PSNI) is established under the Police (Northern Ireland) Act 2000. The Police (Northern Ireland) Act 2000 is structured in a similar way to the Police Reform and Social Responsibility Act 2011 in England and Wales, with many similar provisions. However, there is no directly elected policing in Northern Ireland, and the chief constable is held to account for the exercise of their functions—and those of their officers and staff—by the Northern Ireland Policing Board (NIPB). The Board is an independent public body made up of ten political and nine independent members, established to ensure for all the people of Northern Ireland an effective, efficient, impartial, representative, and accountable police service, which will secure the confidence of the whole community by reducing crime and the fear of crime. Members of the Board are bound by a code of conduct and operate under agreed standing orders.

The NIPB has a statutory duty to monitor the National Crime Agency's (NCA) functions in Northern Ireland. While the NCA operates UK-wide against serious crime, its activities in Northern Ireland are overseen by the NIPB through biannual meetings, performance reports, and a Memorandum of Understanding. The NCA provides quarterly performance narrative reports on its activities within Northern Ireland to the NIPB. The NIPB ensures that the actions of the NCA in Northern Ireland are transparent and aligned with the community's accountability needs, as detailed in the NCA Framework Document which ensures the NCA's actions in Northern Ireland are scrutinized by the same body responsible for policing oversight (the NIPB), providing a crucial link between national law enforcement and local democratic accountability.

Other statutory bodies overseeing policing include the Police Ombudsman, the Criminal Justice Inspectorate NI, HM Inspectorate of Constabulary, Fire & Rescue Services and the Northern Ireland Audit Office. The Justice Committee oversees the work of the Minister of Justice.

The Police Ombudsman was established under the Police (Northern Ireland) Act 1998 and has a wide remit to investigate complaints about the police (including the Ministry of Defence Police in Northern Ireland) and certain UK Border Force staff. Other relevant legal bodies overseeing the activities of the police in Northern Ireland include the Northern Ireland Human Rights Commission and the Parades Commission; their statutory functions are set out in the relevant legislation. The Police Ombudsman is a 'corporation sole' appointed by the Executive Office, ensuring independence from both the PSNI and the Department of Justice.

The above statutory accountability arrangements and the legal bodies responsible for their functioning play crucial roles in holding police accountable within those jurisdictions of the UK and some of the policing activity will be carried out by the non-geographic police services discussed earlier in the chapter. While the specific framework for each jurisdiction is clearly set out in the legal instruments, complexities can arise where police officers and staff operate in collaboration or under mutual aid provisions. In such operations, the potential for differing laws and policies on AI usage by the police will need to be taken into account, perhaps most importantly where there are significant differences in public attitudes towards AI. Such differences have been clearly evidenced in a national citizen consultation, which is set out in detail in Chap. 10.

Despite any nuances across jurisdiction, there is a strong connection between elected bodies and the population for policing across the UK. This connection ensures that accountability remains a central focus in UK policing, even within the diverse jurisdictional landscape described above.

## 3.8 Harmonisation with the EU AI Act

Since the EU's *Harmonisation of Rules on Artificial Intelligence* (Proposal 0106/2021), the policy debate as to how to balance the potential benefits against harms has dominated the area of legal regulation across EU member states. Ultimately the EU AI Act (Regulation (EU) 2024/1689) was passed and lays down harmonised rules on Artificial Intelligence. It was claimed as the first comprehensive legal framework on AI worldwide, and its aim is to 'foster trustworthy AI in Europe'.[25]

The details of the Act and its relevance to AI accountability are set out in Chap. 4. For the purpose of this chapter, it should be noted that the EU AI Act takes a risk-based approach to regulation of AI systems, tools, and deployments. In particular, it places strict legal requirements on some AI capabilities and use cases that are regarded as representing significant risks. These ***high-risk*** use-cases include:

[25] https://digital-strategy.ec.europa.eu/en/policies/regulatory-framework-ai

- AI use-cases in law enforcement that may interfere with people's fundamental rights (e.g. evaluation of the reliability of evidence).
- AI systems used for remote biometric identification, emotion recognition and biometric categorisation (e.g. facial recognition and matching).
- AI use-cases in migration, asylum and border control management (e.g. automated examination of visa applications).
- AI solutions used in the administration of justice and democratic processes (e.g. AI solutions to prepare court rulings).

The type of obligations on users of AI include:

- Making adequate risk assessment and mitigation systems.
- Ensuring high-quality datasets feeding the system to minimise risks of discriminatory outcomes.
- Logging of activity to ensure traceability of results.
- Detailed documentation providing all information necessary on the system and its purpose for authorities to assess its compliance.
- Clear and adequate information to the deployer.
- Appropriate human oversight measures.
- High level of robustness, cybersecurity and accuracy.

Many of these requirements are reflected in the AIPAS principles as discussed in Chap. 5, but it is important to note that the principles are *not* a complete match for the legal provisions applicable in EU member states, and it is essential to check the source legislation and guidance when working with policing organisations in those jurisdictions such as Europol.

In March 2024, the UK government introduced the Artificial Intelligence (Regulation) Bill provisions surrounding the function of AI in the United Kingdom, including a clause to introduce an AI Authority which would create a direct regulatory accountability mechanism. The Bill subsequently lapsed as a result of the general election later that year. As a result, unlike EU member states, the UK currently has no statutory AI regulatory framework setting specific requirements for the policing and security community.

## 3.9 Conclusion

AI-enabled capability is infinitely multi-functional—that is one of its most exciting strengths. Given the powers and privileges of the police and the state, the use of AI in the context of policing and law enforcement brings specific challenges and responsibilities, and these require specific areas of scrutiny, transparency, and assurance.

As the police can do more with AI technology, it will be increasingly important to show what they are *not* doing with it; so important that their use of AI may become a proxy measure for trust and confidence in the police generally. And if the citizen does not trust their police with new technology, the problem is not the technology.

The law regulates the actions of the police and the state generally, together with the use of data and technology by everyone. However, legislative frameworks only go so far and vary significantly across international borders, as do societal expectations (see Chap. 10). The AI accountability challenge for policing and law enforcement will be to balance three things: the technological (what can be done), the legal (what must/must not be done), and the societal (what people will support/expect to be done).

The purpose of the AI accountability principles and contextual discussions set out in the remaining chapters is to help the police, and those holding them to account, to achieve that dynamic balancing of innovation, regulation, and expectation.

## References


1. B. Akhgar, P.S. Bayerl, K. Bailey, R. Dennis, S. Heyes, A. Lyle, A. Raven, F. Sampson, M. Gercke, *AP4AI Framework Blueprint* (2022). https://www.ap4ai.eu
2. H.S. Antunes, P.M. Freitas, A.L. Oliveira, C.M. Pereira, E. Vaz de Sequeira, L.B. Xavier, *Multidisciplinary Perspectives on Artificial Intelligence and Law* (Spring, 2023). https://doi.org/10.1007/978-3-031-41264-6
3. Christie, L. (2021). AI in Policing and Security. Post: https://post.parliament.uk/ai-in-policing-and-security/
4. Department for Science, Innovation and Technology, *Introduction to AI Assurance* (United Kingdom Government, 2024) https://assets.publishing.service.gov.uk/media/65c64a1b14b83c000ea7164d/Introduction_to_AI_Assurance.pdf
5. G.M. Dickinson, Law proofing the future. Harvard J. Legislation **63**(1), 81–136 (2026)

6. European Commission for Democracy (Venice Commission), *Rule of Law Checklist. CDL-AD(2016)007* (Council of Europe, 2016)
7. G. Gabrielli, The use of facial recognition Technologies in the Context of peaceful protest: the risk of mass surveillance and the implications for protection of human rights. Eur. J. Risk Regulat. **16**(2), 514–541 (2025)
8. A. Janjeva, A. Harris, S. Mercer, A. Kasprzyk, A. Gausen, *The Rapid Rise of Generative AI: Assessing Risks to Safety and Security* (CETaS Research Reports, 2023)
9. I. Loader, Revisiting the police Mission, in *Insight Paper no 2, Strategic Review of Policing* (The Police Foundation, 2020)
10. N. Polo, Ohrvik-Stott, *An Eye on the Future* (Ada Lovelace Institute, 2025)
11. F. Sampson, AI is coming to policing—Accountability is coming with it. Biometric Update (2024, November 4). https://www.biometricupdate.com/202411/ai-is-coming-to-policing-accountability-is-coming-with-it

**Fraser Sampson** LL.B. (Hons.), LL.M., MBA., Ph.D., Solicitor is Professor of Governance and National Security at CENTRIC., Sheffield, United Kingdom He was the UK's first combined Biometric and Surveillance Camera Commissioner from 2021 to 23 and is currently Strategic Advisor to the Centre for Emerging Technology and Security at the Alan Turing Institute , London, United Kingdomand a member of the CENTRIC Advisory Board.

**John McNeill** BSc., LL.M., M.Phil. has extensive experience of criminal justice across the UK. He has informed the governance, accountability, operational practices, and ethical activities of a range of professional bodies, including Probation, Prisons, Parole, Policing, and the Armed Forces. He has held numerous Honorary and Visiting Professorships, and Ministerial and Parliamentary appointments.

CHAPTER 4

# Artificial Intelligence in Law Enforcement Operations: The EU AI Act as the Governing Legal Standard

*Marco Gercke*

## 4.1 Introduction

On 1 August 2024, Regulation (EU) 2024/1689, the Artificial Intelligence Act (AI Act), entered into force. It is the world's first comprehensive binding legal framework specifically governing Artificial Intelligence systems. No equivalent instrument of comparable scope and binding force currently exists at the international level. While the Council of Europe's Framework Convention on Artificial Intelligence, Human Rights, Democracy and the Rule of Law, opened for signature in September 2024, represents a parallel international-law instrument, the AI Act constitutes the primary operational regulatory framework governing AI systems within the European Union. For Law Enforcement Agencies (LEAs)s operating in the European Union or deploying

M. Gercke (✉)
Cybercrime Research Institute, Cologne, Germany
e-mail: gercke@cybercrime.de

 
B. Akhgar and P. S. Bayerl (eds.), *Accountability for Artificial Intelligence in Policing and Security*, Security Informatics and Law Enforcement, https://doi.org/10.1007/978-3-032-41284-3_4

AI tools, whose outputs affect persons within EU territory, the AI Act is therefore the central operative legal standard.

The AI Act has significant implications for the work of LEAs. The Regulation imposes binding obligations on LEAs as deployers of certain AI systems, prohibits specified investigative practices outright, and establishes requirements concerning human oversight, logging, transparency, and data governance. In addition, it requires a Fundamental Rights Impact Assessment (FRIA) before the deployment of certain high-risk AI systems used in law enforcement. Violations may lead to administrative sanctions. The AI Act thus marks an important shift in the accountability framework governing LEA use of AI, transforming what was previously largely an ethics-based governance discourse into enforceable legal obligations [9, 10].

The territorial reach of the AI Act is not limited to providers established within the European Union. Providers established outside the EU remain subject to the Regulation where AI system outputs are used within the Union. This extraterritorial mechanism resembles the approach adopted by the General Data Protection Regulation (GDPR) and exemplifies what Bradford [3] terms the 'Brussels Effect': by conditioning access to the European market on compliance with EU regulatory standards, the Union extends the global influence of its regulatory model through economic necessity rather than direct legal compulsion [2]. In practice, this means that the AI Act is likely to influence the design and availability of AI systems used by LEAs worldwide, irrespective of their domestic regulatory environment.

The drafters of the AI Act explicitly addressed law enforcement uses of AI. The Regulation contains dedicated provisions on prohibited AI practices with specific law enforcement exceptions, classifies many law enforcement AI applications as high-risk, and establishes transparency obligations that are modified in certain law enforcement contexts.

Section 4.2 in this chapter traces the development of the AI Act from its political origins through the trilogue negotiations, focusing on the law enforcement dimensions of that process. Section 4.3 provides a structured analysis of the substantive provisions most directly affecting LEA operations, including prohibited practices, high-risk classifications, deployer obligations, transparency requirements, data protection interfaces, governance structures, and individual rights. Section 4.4 concludes with observations on compliance strategy.

## 4.2 The Development of the AI Act

### 4.2.1 *From Ethics to Legislation: 2018–2021*

The binding regulation of artificial intelligence at Union level emerged from a gradual soft-law process. In April 2018, the European Commission published its Communication 'Artificial Intelligence for Europe', identifying excellence and trust as twin policy objectives [9]. The Commission subsequently convened the High-Level Expert Group on Artificial Intelligence (AI HLEG), whose Ethics Guidelines for Trustworthy AI, published in April 2019, translated those objectives into seven operational requirements:

1. Human agency and oversight
2. Technical robustness and safety
3. Privacy and data governance
4. Transparency
5. Diversity, non-discrimination and fairness
6. Societal and environmental well-being
7. Accountability

While these guidelines became influential for LEAs and for EU-funded research projects developing AI tools for law enforcement purposes, they lacked binding legal force. Nevertheless, they provided much of the normative foundation upon which the later legislative framework was built.

Compared with the Ethics Guidelines for Trustworthy AI, the AI Act adopts a narrower and more operationalised understanding of trustworthiness. As Laux et al. [10] observe, the Regulation largely conceptualises trustworthiness in terms of the acceptability and management of AI-related risks through conformity assessment and regulatory compliance mechanisms. This conflation of 'trustworthiness' with 'risk acceptability', rather than with broader public trust, may not necessarily generate the societal confidence that the Regulation seeks to promote. For law enforcement decision makers, this critique highlights an important practical point: compliance with the AI Act does not automatically produce public trust in AI-assisted policing.

The Commission's White Paper on Artificial Intelligence (February 2020) translated the ethics-based framework into a legislative agenda, proposing a dual-pillar approach combining innovation support with proportionate regulation of high-risk AI systems [9].

### 4.2.2 *The Commission Proposal and the Law Enforcement Debate*

On 21 April 2021, the Commission presented its proposal for the AI Act. Veale and Borgesius [19] characterised the proposal as one that 'mixes reduction of trade barriers with broad fundamental rights concerns in a structure unfamiliar to many information lawyers' (p. 98). The proposal introduced the now-familiar risk-based architecture distinguishing between prohibited, high-risk, limited-risk, and minimal-risk AI uses. For LEAs, the proposal already contained a qualified prohibition on certain forms of real-time remote biometric identification in publicly accessible spaces, restrictions on predictive policing practices based solely on profiling, and the classification of various law enforcement AI applications as high-risk under Annex III [16].

Smuha et al. [17] identified an early structural weakness in the proposal: unlike Annex III, which the Commission could amend through delegated acts, the list of prohibited AI practices in Article 5 was comparatively static. The prohibited-practices regime, they argued, appeared heavily influenced by contemporary controversies such as social scoring and remote biometric identification by law enforcement authorities, despite the difficulty of predicting future AI uses. This critique proved prescient, as the rapid emergence of general-purpose AI and large language models during 2022 and 2023 required substantial additions to the regulatory framework rather than adaptation of the original prohibited-practices structure.

The law enforcement community actively participated in the legislative debate. In May 2022, European police chiefs issued a joint declaration acknowledging that AI systems are particularly well suited to supporting investigative activities, whilst cautioning against regulatory approaches that might unduly impede legitimate law enforcement functions [16, p. 601]. Their intervention reflected a central tension underlying the legislative process: the balance between fundamental rights protection and operational effectiveness in policing.

### 4.2.3 *Council, Parliament, and Trilogue*

The Council of the European Union adopted its General Approach in December 2022, refining several law enforcement provisions and modifying the conditions governing exceptions for real-time remote biometric identification. In June 2023, the European Parliament adopted its position at first reading, proposing a substantially more restrictive approach to law enforcement uses of AI than that reflected in the Council text.

The trilogue negotiations concluded in December 2023. The resulting compromise tightened the conditions governing real-time remote biometric identification by introducing stricter authorisation requirements, mandatory Fundamental Rights Impact Assessments (FRIAs), and notification obligations, whilst maintaining a comparatively more permissive framework for certain forms of post-remote biometric identification. This distinction, discussed further in Sect. 4.3.3.3, has attracted sustained criticism in the academic literature. The final text was published in the Official Journal on 12 July 2024 as Regulation (EU) 2024/1689. The AI Act and the Council of Europe Framework Convention on Artificial Intelligence, Human Rights, Democracy and the Rule of Law were finalised within the same broader regulatory period in 2024.

### 4.2.4 *Entry into Force and Phased Applicability*

The phased applicability structure of the AI Act is operationally significant for LEAs. The prohibited AI practices contained in Chapter II became applicable on 2 February 2025, meaning that AI systems deployed by LEAs after that date must comply with Article 5 AI Act. Governance provisions and most rules concerning penalties became applicable on 2 August 2025. Most obligations concerning high-risk AI systems, including deployer obligations under Article 26, are scheduled to become applicable from 2026 onwards under the phased implementation regime established by Article 113 AI Act.

### 4.2.5 *The Digital Omnibus Package and Implementation Adjustments*

The AI Act's implementation timeline has recently been revised. On 7 May 2026, the Council Presidency and the European Parliament announced a provisional agreement under the so-called Omnibus VII package to streamline certain rules on AI. The agreement adjusts the timeline for applying rules on high-risk AI systems by up to 16 months, conditional on the Commission confirming that the necessary harmonized standards and tools are available. The new application dates are 2 December 2027 for stand-alone high-risk AI systems under Annex III and 2 August 2028 for high-risk AI systems embedded in products. For LEAs, this means that the most consequential obligations (including deployer requirements under Article 26 and the FRIA under Article 27) are unlikely to acquire full enforceability before late 2027. The agreement also introduces a new prohibition under Article 5 AI Act on the use of AI systems to generate non-consensual sexual content or child sexual abuse material (CSAM), applicable without exception to all deployers including LEAs. Formal adoption is expected in summer 2026.

## 4.3 Relevant Provisions of the AI Act for Law Enforcement

### 4.3.1 *Scope: The Definition of AI System*

The AI Act applies not to artificial intelligence in the abstract, but to 'AI systems' and to general-purpose AI (GPAI) models within the meaning of the Regulation. Article 3(1) defines an 'AI system' as a machine-based system designed to operate with varying levels of autonomy that infers, from inputs, how to generate outputs (including predictions, content, recommendations, or decisions) that can influence physical or virtual environments. A core component of the definition is the system's capacity to infer from inputs how to generate outputs through machine-based processing techniques, including learning, reasoning, or modelling approaches that exhibit a degree of autonomy beyond purely manual human operation.

For LEAs, this definitional threshold creates practical challenges. The AI Act does not generally require providers to identify every software product explicitly as containing an AI system within the meaning of the Regulation. A law enforcement agency acquiring a complex investigative tool may therefore not know with certainty whether it contains one or more AI systems within the meaning of the AI Act. As a consequence, procurement specifications should require explicit contractual representations from software providers concerning AI Act applicability. In cases of doubt, LEAs should seek independent technical assessment.

It is also worth noting that the AI Act adopts a broader conception of an AI system's lifecycle than conventional product safety laws. Regulatory obligations begin at the design and development stages instead of the time a product is placement on the market. This provides important context for the documentation and traceability requirements established in Articles 11 to 13 AI Act. The intent of those documentation requirements is to ensure that compliance-relevant decisions made during development are captured and are auditable throughout the system's operational life.

### 4.3.2 *The Risk-Based Approach and Its Critiques*

The AI Act's operative mechanism is a risk-based classification distributing regulatory obligations in proportion to assessed risk. Hohmann and Kollar [8, p. 1] characterise the architecture as a risk-based framework distinguishing between prohibited, high-risk, limited-risk, and minimal-risk AI uses, distributing regulatory burdens in order to address AI-related risks 'while ensuring the enforcement of the public interest and a high level of protection rooted in European cultural and political foundations'. For LEAs, the challenge is that the categories of AI application most valuable to operational policing (biometric identification, predictive analytics, and suspect profiling) are precisely those subject to the highest regulatory burdens.

The risk-based approach has attracted critique. Rangone and Megale [15, p. 1082] observe that the AI Act's ex ante risk-based approach 'might lead to regulatory obsolescence' given the pace of AI development and risks 'over or under-inclusion of AI applications in risk categories'. Specifically in the law enforcement context, Smuha et al. [17, p. 32] noted early that Annex III's law enforcement category, by focusing on systems involving natural persons as subjects, risks omitting geospatial crime analytics—area-based predictive policing systems that do not profile identifiable individuals—from high-risk classification despite their potential for discriminatory resource allocation.

The Commission has since published Draft Guidelines on the classification of high-risk AI systems under Article 6(3), which provide clarification of the exemption criteria. The Draft Guidelines point out that Annex III categories must be interpreted in light of the intended purpose of the AI system and its actual impact on decision-making processes. A system that merely assists a human decision without materially influencing it may fall outside the high-risk classification even where it nominally falls within an

Annex III category. This distinction has practical relevance for LEAs operating analytical tools in an advisory rather than determinative capacity. However, it should be noted that these are currently only guidelines and changes could happen.

### 4.3.3 *Prohibited AI Practices: Article 5*

#### 4.3.3.1 *Prohibitions Without Law Enforcement Exception*

Article 5 AI Act identifies a number of AI practices as presenting unacceptable risks and prohibits them outright. Several prohibitions apply without any exception whatsoever, including to law enforcement. The prohibition most relevant for LEAs is Article 5(1)(e) AI Act, which prohibits the use of AI systems that create or expand facial recognition databases through untargeted scraping of facial images from the internet or CCTV footage. This prohibition is absolute. It directly constrains the methodological approach of several commercially available investigative tools, including Clearview AI, that have been used by LEAs across Europe and internationally [17, p. 7].

#### 4.3.3.2 *Individual-Based Predictive Policing*

Article 5(1)(d) AI Act prohibits AI systems used to assess the risk of an individual committing a criminal offence based solely on profiling or on personality traits and characteristics. The prohibition does not extend to AI systems supporting a human assessment of individual criminal involvement where that assessment is already grounded in objective and verifiable facts directly linked to criminal activity. Recital 42 makes the legislative intent explicit: the prohibition seeks to preserve the presumption of innocence by ensuring that individuals are judged on actual conduct rather than algorithmic prediction alone.

The prohibition does not extend to area-based predictive policing systems operating through the assessment of crime probabilities at specific locations using historical or statistical data rather than individual profiling. Systems such as PredPol or CompStat, which identify crime hotspots by correlating location, time, and historical offence patterns, therefore fall outside Article 5(1)(d). Recital 42 further clarifies that risk analytics used to assess the likelihood of financial fraud or the localisation of narcotics on the basis of suspicious transaction patterns or known trafficking routes are likewise not covered where they are not based solely on profiling of identifiable individuals.

#### *4.3.3.3 Biometric Categorisation*

Article 5(1)(g) AI Act prohibits the use of biometric categorisation systems to deduce or infer an individual's race, political opinions, trade union membership, religious or philosophical beliefs, sex life, or sexual orientation. The prohibition does not, however, prevent certain forms of technical labelling or filtering of lawfully acquired biometric datasets in the law enforcement context. Recital 30 gives the example of categorisation based on objectively observable physical features such as hair colour or eye colour. However, the prohibition continues to apply to the deduction or inference of sensitive characteristics such as political opinions, religious beliefs, or sexual orientation from biometric data.

#### *4.3.3.4 Real-Time Remote Biometric Identification in Publicly Accessible Spaces*

This prohibition is among the most operationally consequential provisions of the AI Act for policing. Article 5(1)(h) prohibits LEAs from deploying 'real-time' remote biometric identification systems in publicly accessible spaces. A remote biometric identification system is defined as an AI system identifying natural persons through comparison of their biometric data against a reference database. Identification is considered 'real-time' where there is no significant delay between data capture, comparison, and identification. 'Publicly accessible spaces' refers to physical spaces accessible to an indeterminate number of persons. Online environments are expressly excluded from this definition.

The prohibition admits four narrowly framed exceptions; all subject to a strict necessity requirement. Each use requires prior authorisation by a judicial authority or by an independent administrative authority whose decision is binding, subject in urgent cases to retrospective authorisation within 24 hours. Deployment of such systems is further subject to the FRIA requirements of Article 27 AI Act, as well as notification obligations vis-à-vis the market surveillance authority and the national data protection authority.

Permitted uses are limited to:

- The targeted search for victims of abduction, trafficking, or sexual exploitation, and missing persons
- The prevention of a specific, substantial, and imminent threat to life or of a terrorist attack

- The identification of suspects for the investigation, prosecution, or execution of criminal penalties for Annex II offences punishable by a maximum custodial sentence of at least four years

The exclusion of online spaces from the prohibition creates a significant interpretive gap. Article 5(1)(h), read together with Article 3(44), produces the anomalous result that real-time facial recognition on a live CCTV feed in a railway station is prohibited, subject to exceptions, whereas the same identification performed on a live internet video stream falls outside the prohibition. Such online identification may nevertheless qualify as a high-risk AI system under Annex III No. 1(a), although without being subject to the stricter safeguards of Article 5. Recital 95 sentence 4 AI Act explicitly states that the conditions governing post-remote biometric identification must not be used to circumvent the stricter conditions applicable to real-time identification systems. A LEA systematically routing biometric identification through online streams in order to avoid Article 5(1)(h) would therefore engage directly with this anti-circumvention principle.

### 4.3.4 *High-Risk AI Systems*

#### 4.3.4.1 *Classification*

Chapter III of the AI Act establishes the regulatory framework for AI systems posing significant risks to health, safety, or fundamental rights. For LEAs, the relevant classification mechanism is Article 6(2) in connection with Annex III. Annex III identifies several categories of stand-alone high-risk AI systems, including systems used in biometrics (No. 1) and law enforcement (No. 6), both of which are of direct operational relevance to LEAs.

Classification under Annex III does not automatically result in high-risk status in every case. Based on Article 6(3) AI Act, an Annex III system is not to be considered high-risk where it does not pose a significant risk of harm to the health, safety, or fundamental rights of natural persons, including where the system merely performs a narrow procedural or preparatory task, improves the result of a previously completed human activity, or otherwise does not materially influence decision-making in the relevant area.

Annex III No. 6 lists five use-cases applicable to LEAs and agencies acting on their behalf, classified as high-risk 'because their accuracy, reliability and transparency is particularly important to avoid adverse impacts, retain public trust and ensure accountability and effective redress'. The five use-cases are:

1. AI systems assessing the risk of an individual becoming a victim of criminal offences (area-based predictive policing)
2. AI systems used as polygraphs or similar credibility assessment tools
3. AI systems evaluating the reliability of evidence in criminal investigations or prosecutions
4. AI systems assessing the risk of an individual offending or re-offending, or assessing personality traits and past criminal behaviour
5. AI systems used for the profiling of individuals in the detection, investigation, or prosecution of criminal offences

This taxonomy covers a wider range of AI capabilities that LEAs currently deploy (see example use cases in Chaps. 12, 13, and 14).

*4.3.4.2 Mandatory Requirements*

A high-risk AI system must satisfy a set of mandatory requirements before it can be placed on the market or put into service. These requirements, established in Arts. 9–15 AI Act, include:

1. The establishment and maintenance of a risk management system continuously identifying and evaluating risks throughout the system's lifecycle
2. Compliance with data and data governance requirements, including the use of training, validation, and testing datasets meeting standards of relevance, representativeness, and, as far as possible, freedom from errors and completeness
3. The preparation of appropriate technical documentation
4. The implementation of automatic logging capabilities ensuring an appropriate level of traceability
5. The provision of instructions for use and other information to deployers
6. The design and implementation of effective human oversight measures

7. The achievement of appropriate levels of accuracy, robustness, and cybersecurity

Compliance with these requirements must generally be demonstrated through a conformity assessment procedure. Following successful assessment, the provider draws up an EU declaration of conformity and affixes the CE marking before the system may be placed on the market or put into service.

The dataset quality requirement demands separate attention. Article 10(2)(f) AI Act explicitly requires that training, validation, and testing datasets be examined for possible biases that are likely to affect health and safety or lead to discrimination prohibited under Union law. This provision directly confronts a documented structural problem in law enforcement AI. Historical crime data reflects pre-existing patterns of enforcement activity that may themselves be discriminatory. A biometric identification system trained predominantly on one demographic will generate systematically higher false positive rates for underrepresented groups. Researchers have documented this differential impact extensively [4, 11]. The AI Act does not resolve the underlying structural problem, but it creates clear legal exposure for providers who deploy systems with demonstrably unrepresentative datasets and for LEAs that operate such systems without adequate human oversight and documentation. As Caseburo and March (2024) note, 'if the AI-based unfavourable treatment relies on prohibited grounds and is not adequately justified, law enforcement AI can violate the rights to equality and non-discrimination as safeguarded by Articles 20 and 21 of the Charter' (p. 1571).

#### *4.3.4.3 Deployer Obligations*

Based on Article 26 AI Act, an LEA deploying a high-risk AI system must take appropriate technical and organisational measures to ensure that the system is used in accordance with the provider's instructions for use; assign human oversight to persons possessing the necessary competence, training, and authority; ensure that input data is relevant and sufficiently representative in view of the system's intended purpose; monitor the operation of the system on the basis of the instructions for use and inform the provider and relevant authorities where risks within the meaning of Article 79 AI Act arise; suspend use where the system presents a risk to health, safety, or fundamental rights; and retain automatically generated logs to the extent such logs are under the deployer's control, for a period appropriate to the intended purpose of the system and, for high-risk AI systems used in law enforcement, for at least six months unless Union or national law provides otherwise.

In addition, LEAs deploying high-risk AI systems listed in Annex III are generally required to conduct a Fundamental Rights Impact Assessment pursuant to Article 27 AI Act. Where the deployment involves post-remote biometric identification systems in the context of targeted law enforcement investigations, additional conditions under Article 26(10) AI Act apply, including, in principle, prior authorisation by a judicial authority or by an independent administrative authority whose decision is binding.

#### *4.3.4.4 Fundamental Rights Impact Assessment*

Public bodies deploying high-risk AI systems referred to in Annex III (including, in most cases, LEAs) are required under Article 27 AI Act to conduct a Fundamental Rights Impact Assessment (FRIA) before putting the system into use. The FRIA must include, inter alia, a description of the processes in which the system will be used; the period and frequency of intended use; the categories of natural persons and groups likely to be affected; the specific risks of harm to fundamental rights; the human oversight measures to be implemented; and the arrangements for governance, supervision, and complaint handling. Following completion of the FRIA, the deployer must notify the relevant market surveillance authority that the assessment has been conducted and provide the information specified in Article 27 AI Act.

### *4.3.5 Transparency Obligations*

Article 50 AI Act establishes several transparency obligations that are relevant for LEAs because they contain specific law enforcement exemptions. Providers of AI systems interacting directly with individuals must generally disclose that the interaction involves AI, unless the use is authorised by law for the detection, prevention, investigation, or prosecution of criminal offences. Similar exemptions apply to certain obligations concerning AI-generated or manipulated content and to the use of emotion recognition or biometric categorisation systems. These exemptions create operational space for covert or non-transparent AI use in criminal investigations. They do not, however, remove the obligation to comply with the Law Enforcement Directive, the Charter of Fundamental Rights, and applicable national procedural rules governing covert investigative measures.

The academic literature on AI transparency provides important context for understanding the practical significance of these obligations. Felzmann et al. [6, p. 3333] conceptualise 'transparency by design' as a requirement that AI systems be designed so that decision-making processes can be retraced, with records allowing reliable reconstruction of relevant decisions and attribution of responsibility. This concept of traceability is reflected in the AI Act's requirements concerning logging, record-keeping, technical documentation, and information provision to deployers, particularly in Articles 11–13 AI Act. As noted in the transparency literature mapping the EU regulatory framework for AI, design flaws in algorithms or training datasets can cause failures that lead to violations of fundamental rights, yet 'some degree of opacity can be systemic and resistant to attempts at transparency, especially when the use of machine learning algorithms makes deductive explanation impossible' [5, p.10].

### 4.3.6 The Data Protection Interface

The AI Act does not operate in isolation. It coexists with the Law Enforcement Directive (LED) and, outside the law enforcement context, the GDPR. The LED, as the primary data protection instrument governing police and criminal justice processing, applies to personal data processed by AI systems deployed by LEAs. Article 11(1) LED establishes that decisions based solely on automated processing, including profiling, which produce an adverse legal effect concerning the data subject or significantly affect the data subject, are prohibited unless authorised by Union or Member State law providing appropriate safeguards, at a minimum including the right to obtain human intervention. This provision interacts closely with the AI Act's restrictions on predictive policing and other high-risk law enforcement AI uses, both instruments requiring meaningful human involvement where automated systems substantially influence decisions affecting individuals.

Hohmann and Kollár [8, p. 1] note that the AI Act 'supplements the GDPR through a risk-based approach that subjects high-risk AI systems to specific obligations' and that its central concerns include 'algorithmic opacity, bias, and legal uncertainty in profiling and automated decision-making'. Their analysis illustrates how the AI Act seeks to mitigate these risks through strengthened requirements concerning risk management, human oversight, transparency, and data governance. The FRIA and DPIA obligations are parallel but distinct: the FRIA focuses broadly on risks to fundamental rights, whereas the DPIA focuses specifically on risks arising from personal data processing. For LEAs deploying biometric identification or predictive analytics systems, both assessments will often be required.

### *4.3.7 Governance and Enforcement*

The AI Act establishes a complex governance architecture combining EU-level coordination with decentralised national enforcement. At the EU level, the AI Office plays a central role in the implementation, supervision, and coordination of the AI Act, particularly regarding general-purpose AI models. At national level, EU Member States must designate competent authorities, including at least one market surveillance authority responsible for supervision and enforcement within their territory. Novelli et al. [13, p. 590] describe this governance structure as 'intricate, yet solid', comprising the AI Office, the European AI Board, the Advisory Forum, the Scientific Panel, and national competent authorities whose effective functioning depends on close interinstitutional coordination (for a UK perspective see Chaps. 2 and 3).

The enforcement framework contains specific safeguards for law enforcement AI systems. Where LEAs deploy high-risk AI systems referred to in Annex III points 1, 6, or 7, the technical documentation must remain within the premises of those authorities. Market surveillance authorities may obtain access upon request, but access is restricted to personnel holding the required security clearance. Söderlund and Larsson [18, p. 40] characterise this structure as reflecting the AI Act's decentralised enforcement model, under which EU Member States retain primary responsibility for operational supervision while Market Surveillance Authorities (MSA) are granted extensive investigative powers, including access to training datasets and, where necessary and proportionate, source code.

### 4.3.8 *Individual Rights and the Accountability Gap*

One aspect relevant in the broader context of enforcement is the absence of general substantive individual rights enforceable against providers or deployers. Smuha et al. [17, p. 51] identify this as a significant weakness in AI Act and point out that the lack of substantive individual rights in the AI Act 'reduces individuals to entirely passive entities, unacknowledged and unaddressed in the regulatory framework'. The AI Act operates primarily through obligations imposed on providers and deployers, enforced by regulatory authorities and not rights exercised by individuals against AI operators.

Rangone and Megale [15, p. 1082] extend this analysis, observing that 'the European values foreseen in the Act are likely to remain primarily aspirational' and that the Act risks 'neglecting adequate protection of central aspects intertwined with the development of AI systems'. Their observation has direct operational relevance for LEAs. Compliance with the AI Act's procedural and technical requirements does not immunise a LEA from challenge under the ECHR, the EU Charter, the Law Enforcement Directive, or national law. The Bridges litigation in England and Wales (*Bridges v Chief Constable of South Wales Police*), where facial recognition use by the South Wales Police was found unlawful, not because of a specific AI regulation but through the application of existing data protection and equality legislation, illustrates that compliance with one regulatory framework (the AI Act) does not guarantee lawfulness across all applicable frameworks. As a consequence, LEAs should embed AI Act compliance in a broader compliance regime that includes ECHR, the EU Charter, the Law Enforcement Directive and national law (see also Chap. 5).

Experience from EU funded projects in the FCT (Fight against Crime and Terrorism) domain underlines that building and maintaining public trust in AI-assisted policing requires engagement that goes beyond legal compliance. The academic literature on trust in AI in the public sector consistently identifies transparency, explainability, and meaningful human oversight as the functional prerequisites for citizen acceptance ([6]; also Chap. 10). Laux et al. [10, p. 3] observe that 'even allegedly trustworthy persons, institutions, and technologies might not be trusted after all' by populations who have historically faced discrimination, whose distrust of broadly accepted signals of trustworthiness is rational.

## 4.4 Conclusion

The EU AI Act is a binding EU law with implications for LEAs. Three conclusions emerge from the foregoing analysis. First, the AI Act ends the era in which, in some cases, law enforcement could have deploy AI tools without explicit legal authority, structured accountability, and documented human oversight. The combination of registration requirements, FRIA obligations, mandatory logging, and other limitations constitutes an accountability framework without precedent in prior legislation governing law enforcement.

Second, in areas with operational relevance for LEAs (the prohibition on individual-based predictive policing without independent factual basis, the conditional prohibition on real-time biometric identification, and the high-risk classification of five categories of law enforcement AI application) LEAs are forced to undertake legal assessments before deploying specific AI systems. Due to the complexity of the assessment, it is unlikely that they can be delegated to procurement officers or information technology departments.

Third, it is clear that the impact of the EU AI Act is not limited to the 27 EU Member States. Non-EU LEAs procuring AI tools from EU-market-compliant vendors most likely in the future acquire systems designed to meet the requirements of the AI Act. Whether the AI Act's standards will become the global baseline for law enforcement AI procurement will need to be seen.

The challenge for law enforcement leadership is to avoid confusion while implementing the necessary compliance framework. The Regulation's requirements on transparency, human oversight, data quality, and fundamental rights impact assessment are—though likely not intended—bureaucratic burdens imposed upon LEAs. This said, they are (if embedded in a large compliance framework) the conditions under which AI-assisted policing can maintain the public legitimacy without which LEAs cannot operate. As Akhgar et al. [1] observe in the AP4AI Framework for Accountable AI in the Security Domain, accountability in Artificial Intelligence is ultimately the condition of legitimate authority. That observation applies with particular force to the law enforcement community as it navigates the most significant legal transformation of investigative technology in a generation.

## References


1. B. Akhgar, P.S. Bayerl, K. Bailey, R. Dennis, S. Heyes, A. Lyle, A. Raven, F. Sampson, AP4AI Framework Blueprint (2022). https://www.ap4ai.eu
2. M. Almada, A. Radu, The Brussels side-effect: how the AI Act can reduce the global reach of EU policy. German Law J. **25**, 646–663 (2024). https://doi.org/10.1017/glj.2023.108
3. A. Bradford, The Brussels effect. Northwestern Univ. Law Rev. **107**(1), 1–68 (2012). https://doi.org/10.2139/ssrn.2770634
4. Europol Innovation Lab, AI and policing—the benefits and challenges of artificial intelligence for law enforcement. Europol Innovation Lab Observatory Report (2024)
5. H. Felzmann, E. Fosch-Villaronga, C. Lutz, A. Tamo-Larrieux, Transparency you can trust: transparency requirements for artificial intelligence between legal norms and best practices. Soc. Media Soc. **5**(2), 1–14 (2019). https://doi.org/10.1177/2056305119860502
6. H. Felzmann, E. Fosch-Villaronga, C. Lutz, A. Tamo-Larrieux, Towards transparency by design for artificial intelligence. Sci. Eng. Ethics **26**, 3333–3361 (2020). https://doi.org/10.1007/s11948-020-00276-4
7. High-Level Expert Group on Artificial Intelligence, *Ethics Guidelines for Trustworthy AI* (European Commission, 2019). https://digital-strategy.ec.europa.eu/en/library/ethics-guidelines-trustworthy-ai
8. B. Hohmann, G. Kollár, Reflections on the data protection compliance of AI systems under the EU AI Act. Cogent Soc. Sci. **11**(1), Article 2560654 (2025). https://doi.org/10.1080/23311886.2025.2560654
9. M. Hulok, The EU model of AI governance. Digital Soc. **40**, 530–542 (2025). https://doi.org/10.1007/s44206-025-00215-1
10. J. Laux, S. Wachter, B. Mittelstadt, Trustworthy artificial intelligence and the European Union AI Act: on the conflation of trustworthiness and acceptability of risk. Regul. Gov. **18**, 3–32 (2024). https://doi.org/10.1111/rego.12512
11. J. Mökander, M. Axente, F. Casolari, L. Floridi, Conformity assessments and post-market monitoring: a guide to the role of auditing in the proposed European AI regulation. Minds Mach. **32**, 241–268 (2022). https://doi.org/10.1007/s11023-021-09577-4
12. C. Novelli, F. Casolari, A. Rotolo, M. Taddeo, L. Floridi, AI risk assessment: a scenario-based, proportional methodology for the AI Act. Digital Soc. **3**, 13 (2024). https://doi.org/10.1007/s44206-024-00095-1
13. C. Novelli, P. Hacker, J. Morley, J. Trondal, L. Floridi, A robust governance for the AI Act: AI Office, AI Board, Scientific Panel, and national authorities. Eur. J. Risk Regul. **16**, 566–590 (2025). https://doi.org/10.1017/err.2024.57

14. G. Pavlidis, Regulation 2024/1689 (EU Artificial Intelligence Act): the rights-based approach to technological governance. Review of European and Comparative Law. Online First: 23 Feb 2026 (2026)
15. N. Rangone, L. Megale, Risks without rights? The EU AI Act's approach to AI in law and rule-making. Eur. J. Risk Regul. **16**, 1082–1097 (2025). https://doi.org/10.1017/err.2025.13
16. A. Sachoulidou, Harnessing AI for law enforcement: solutions and boundaries from the AI Act. Eur. J. Polic. Stud. **2026**(1), 3–5 (2026)
17. N. Smuha, E. Ahmed-Rengers, A. Harkens, W. Li, J. MacLaren, R. Piselli, K. Yeung, How the EU can achieve legally trustworthy AI: a response to the European Commission's proposal for an artificial intelligence act. SSRN Working Paper (2021). https://ssrn.com/abstract=3899991
18. K. Söderlund, S. Larsson, Enforcement design patterns in EU law: an analysis of the AI Act. Digital Soc. **3**, 41 (2024). https://doi.org/10.1007/s44206-024-00129-8
19. M. Veale, F.Z. Borgesius, Demystifying the Draft EU Artificial Intelligence Act. CRi Comput. Law Rev. Int. **4**(2021), 97–112 (2021)

**Marco Gercke** is a lawyer, professor, entrepreneur and Director of the Cybercrime Research Institute who advises governments, international organisations and global companies on the legal, strategic and societal implications of emerging technologies, including Artificial Intelligence. Having developed machine-learning solutions for medical diagnostics and AI-driven user interaction systems as an entrepreneur, he combines technical experience with legal and policy expertise in the field of AI governance. His current work focuses on the opportunities, risks and regulatory challenges of Artificial Intelligence, including compliance with the EU AI Act, responsible AI deployment, and the impact of AI on society and public institutions.

CHAPTER 5

# Establishing AI Accountability in Policing and Security

*Petra Saskia Bayerl and Babak Akhgar*

## 5.1 Introduction

Artificial Intelligence (AI) presents policing with important capabilities to increase the speed, accuracy, and range of investigations whilst reducing costs and potentially directing resources more effectively and efficiently. Some AI models provide the potential to solve old crimes by helping investigators to identify cases that contain promising evidence which could be re-examined using new forensic techniques. They also help to prevent new crimes, particularly where technology is being used by the perpetrators, for example, to detect cyber-attacks faster, identify large-scale frauds, or fend-off disinformation campaigns, or detect deep fake. In fact, AI use becomes an imperative given that criminals and crime networks increasingly deploy AI against society and States (Chaps. 13, 14).

P. S. Bayerl (✉) · B. Akhgar
CENTRIC (Centre of Excellence in Terrorism, Resilience, Intelligence and Organised Crime Research), Sheffield, UK
e-mail: p.s.bayerl@shu.ac.uk



B. Akhgar and P. S. Bayerl (eds.), *Accountability for Artificial Intelligence in Policing and Security*, Security Informatics and Law Enforcement, https://doi.org/10.1007/978-3-032-41284-3_5

At the same time, there are understandable concerns in society about the police use of AI that are centred around the infringements of privacy, fundamental freedoms, or even critical thinking (e.g. [8, 12]). These concerns also find expression in an increasing tendency to challenge AI use in the courts and the tightening of regulations on national and international levels. The challenge for police is how to capitalize on AI to safeguard society whilst also assuaging societal concern about their use of advanced technological capabilities.

We argue that AI accountability is an important concept that binds organisations to enforceable obligations and thus ensures actionable procedures to assess, evidence, and enhance responsible AI usage in policing. Beyond the concepts of trustworthiness and explainability, AI accountability imposes obligations on organisations to justify their AI usage, monitor its application, and above all, enforce changes and redress if things go wrong.

Whilst accountability seems an abstract concept, it can be a very practical instrument, not only for the proactive risk assessment and management of AI but also to understand the benefits of AI deployment: '*It helps to think through the why, what, how, who and for whom of AI deployments, and it does so proactively. It supports police forces to: understand risks and strengths before each AI procurement, deployment or change; know up-front what to do if something goes wrong; [and] understand which areas of accountability are achieved or need addressing, and have evidence ready in case they get challenged and have to proof adequate AI procedures*' [5, p. 17].

This chapter offers firstly the conceptual foundation for AI accountability in policing, in that it defines AI accountability as a composite of 12 accountability principles. It further explains how to make the concept practical and assessable through the operationalisation of the principles into eight themes.

## 5.2 Defining AI Accountability in a Policing Context

Accountability is the 'acknowledgement and assumption of responsibility for actions, decisions, and their consequences' [10] and comprises in itself the three aspects of monitoring, justification and enforcement [15]. Accountability thus offers the ability to monitor and enforce obligations towards one or multiple stakeholders, in the understanding that not meeting these obligations will lead to consequences.

The importance of accountability lays in its aim of 'preventing and redressing abuses of power' [15, p. 14]. It has thus at its very core the notion of negotiations across various legitimate interests when it comes to the deployment of AI for policing and security purposes.

*AI Accountability* specifically refers to 'the acknowledgement of an organisation's responsibility, when using AI, to act in accordance with the legitimate expectations of stakeholders and the acceptance of the consequences—legal or otherwise—if they fail to do so' [2].

### 5.2.1 *The Need to Consider the Complete AI Lifecycle*

Often, AI accountability discussions are focused on algorithms (e.g. [9]), the purchase of AI capabilities, or deployment stages [1]. However, accountability does not stop once a system is no longer in use. Whilst the facial recognition systems used by the Metropolitan police to identify persons wanted for previous serious criminal offences or individuals representing a risk of violence at public events may only be in use for some hours or days,[1] consequences may last considerably longer—both in a positive sense such as arrests of wanted offenders and in a negative sense—when decisions may have been wrong infringing on the rights of the individual affected. The Harm Assessment Risk Tool (HART) used by Durham Police in the UK to assess the likelihood of suspects for reoffending was heavily criticised for being flawed and to systematically over-estimate the risk of re-offending using 'racist data profiles' [11; also 7].

AI capabilities leave data, decisions, and impacts, even if they are decommissioned. To ensure a comprehensive risk assessment and management, a core requirement is therefore that police organisations evaluate AI accountability across the full lifecycle of their AI capabilities or systems. The lifecycle begins before deployment, with either design or procurement, and ends not with deployment but with post-deployment to handle the legacy of AI capabilities (see Fig. 5.1). We hence argue that the full lifecycle should be conceptualised in six stages:

1. **Pre-deployment**: Encapsulates all considerations and decisions before an AI system or capability enters operational use. This

[1] https://www.met.police.uk/police-forces/metropolitan-police/areas/about-us/about-the-met/facial-recognition-technology/.

encapsulates disparate 'origins' of the AI system, as either self- or co-developed by a police force or as procured from 'outside', creating two entry points:

a. *Research and design:* the AI capability is created or co-created with direct input from the adopting police force; this includes full consideration for training datasets
b. *Procurement:* the AI capability is brought in (e.g. purchased) from a third-party

2. **Deployment**: Encapsulates all considerations and decisions during the operational deployment of an AI capability

    a. *Operational usage:* running of the AI capability according to a specified purpose
    b. *Revision:* adaptation of the AI capability either in (a) its features or (b) in the purpose of its application

3. **Post-deployment**: Encapsulates all considerations and decisions after the operational deployment of the AI capability has ended

    a. *Migration:* the AI capability is replaced by a different AI capability for the same or closely related purpose and data, decisions, etc. are transferred
    b. *Decommissioning:* the AI capability is retired, and no further AI capability is used going forward

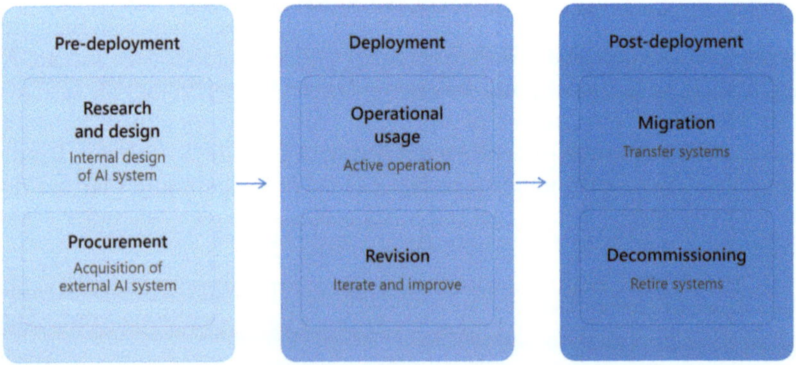


**Fig. 5.1** AI lifecycle stages

This comprehensive view of the AI lifecycle is vital, as each stage is linked to disparate AI accountability considerations, roles and responsibilities. For instance, a police force (co)creating an AI system has considerable influence on the exact design but requires strong internal expertise on algorithms, models, and training data. In contrast, the procurement of a system means that numerous decisions may not be in the hand of the deploying organisations (e.g. training data or algorithms used), which shifts responsibilities for ensuring and evidencing AI Accountability to provider organisations (Chap. 7). Similarly, deployment and post-deployment are likely to involve disparate core functions such as analysts during deployment, and data protection officers during post-deployment.

Since each stage in the AI lifecycle is linked with specific considerations and decision points, they each require stage-specific planning and management (for details see the AIPAS Implementation Guidance in the Appendix). Particularly the transition phases—from pre-deployment to deployment and from deployment to post-deployment—require careful management, when responsibilities, knowledge, data, logs, etc. may need to be handed over to new functions.

### 5.2.2 *AI Accountability as a Process and Network of Mutual Obligations*

The AI lifecycle, whilst important, is still a largely static perspective on accountability, where AI accountability responsibilities and tasks transition (more or less) neatly from one stage to another. Whilst true in the broader sense, in practice accountability is much 'messier'. Our concept of AI accountability is therefore based on two core understandings: that *accountability is a process* and that *it is accomplished in a network of mutual obligations* [5].

#### 5.2.2.1 *Accountability as a Process*

AI deployments require active management to ensure that they remain legal, transparent, responsive to, and in line, with societal expectations. AI accountability is not something that can be assessed once and then assumed. It needs to be assessed and re-assessed across the lifecycle of an AI system. Also, external triggers can require a review: new policies or regulations may change the legal landscape for data collection or retention, new AI capabilities may require a review of decision oversight, etc.

Moreover, accountability obligations such as learning and redress (see Sect. 5.3.1) are by design intended to ensure that accountability arrangements, processes, and mechanisms can adapt to changing circumstances. Transparency, explainability, and inclusivity are further accountability requirements that emphasise process, as communication and negotiation with diverse stakeholder groups is seldom straightforward but needs continuous engagements and refinements.

#### 5.2.2.2 *Accountability as a Network of Mutual Obligations*

Policing accountability implicates a multitude of functions within the police, as well as external stakeholders the police are accountable to (e.g. ministries, mayors, general public) and accountable for (e.g. victims of crime, communities at risk) (Chaps. 2, 3). The introduction of AI does not change this but rather adds additional obligations, such as avoidance of algorithmic biases, automation biases, and human oversight for decisions.

However, we argue that accountability is not only from police to others; instead, there are also obligations that other stakeholders have towards the police. An example are the responsibilities of AI providers to be transparent about training data and faults in algorithms of their systems that police may purchase. Another example is the reliance of policing on the public for support such as the reporting of crimes, information relevant to investigations, or even simply the adherence to laws. As we formulated elsewhere, '*the creation of such relationships may well carry a legitimate expectation on behalf of internal security actors that citizens will attract some degree of accountability for the data they contribute*' [4, p. 4].

Mapping out a comprehensive view of relevant stakeholders for a specific (planned or current) AI deployment is crucial to understanding these mutual obligations. We refer to this comprehensive overview as the 'stakeholder landscape' that allows for the systematic outline of which functions or organisations are accountable to whom, in which way, and what respective requirements and needs are for information, communication, engagement, evidence, etc. This enables the police to target accountability measures in the correct way to ensure all stakeholders are able to understand the purpose, remit, and (potential) impacts of

AI deployments. The objective is to make the AI accountability process practical, so it can be translated into concrete steps and measures.

## 5.3 Making AI Accountability Practical and Assessable

AI accountability is not separate from police accountability; if done right, it is an integral part of police accountability and every-day accountability arrangements.

Police accountability is a broad and complex concept that does not lend itself easily to assessment and quantification. A common approach to make police accountability more accessible is a principles-based approach. Principles have been used in numerous AI ethics frameworks, such as OSCE or ALTAI (Assessment List for Trustworthy AI [13]), where abstract concepts such as responsible AI or trustworthy AI are broken down into their defining components (i.e. principles). ALTAI, for example, decomposes trustworthy AI into seven principles ranging from *Transparency* and *Diversity, non-discrimination and fairness* to *Human Agency and Oversight*, with each principle further specified by questions that aim to assess whether this principle has been fulfilled.

The same approach has also been fruitfully applied to policing accountability in the UK. The framework by Lennon et al. [14] who proposed 12 principles that together create a conceptual basis to establish police accountability.

Using an earlier version of this framework as a foundation, we engaged in a bottom-up process that contextualised police accountability to AI in a policing and security context. The AP4AI (Accountability Principles for AI [2]) project was framed within a European context with the aim to define AI accountability in a way that makes it translatable into practical mechanisms, as well as transferable across deployment and national contexts. The process involved extensive consultations with over 100 subject-matter experts from law enforcements, courts, ethics, industry, and civil society across 28 countries to identify the relevant 'universal' components (i.e. principles) of AI accountability in policing (details of the process can be found in [2, 3]). The result was a set of 12 AP4AI principles, which were further validated in an international citizen consultation with over 6,500 citizens in 30 countries (all 27 EU Member States, UK, US, Australia [6]).

This foundation was further developed in the subsequent UK-based project AIPAS (AI Accountability for Policing and Security)[2] in which the 12 principles were re-validated and refined for the national UK context, as well as four specific policing areas: forensics, counter terrorism, serious and organised crime, and child sexual exploitation (see Chap. 8 for details). Below we detail the final AI accountability principles and their relevance for the AI accountability process.

### 5.3.1 *The 12 AI Accountability (AIPAS) Principles*

The final set of AI accountability principles consists of 12 components that together allow to describe the full scope of obligations and requirements across the AI lifecycle and with explicit consideration of needs by all relevant stakeholders. The following list gives the short definition of the AIPAS principles:

1. **Lawfulness—'follow the law'**: All aspects of the use of AI should be lawful. Lawfulness includes compliance with specific legal requirements such as the EU AI Act (where applicable), as well as organisational and professional policies which must be clearly identified and readily available
2. **Completeness—'leave nothing out'**: AI accountability must cover all relevant aspects of the AI ecosystem and lifecycle including its design, supply, deployment, regulation, monitoring, and impacts
3. **Inclusivity—'leave no one out'**: AI accountability arrangements must involve all relevant stakeholders engaged in and affected by an AI deployment
4. **Transparency—'be open'**: AI accountability needs clear, accurate and meaningful information (subject to operational sensitivities) to enabling those directly and indirectly affected or responsible for oversight to make informed judgments and risk assessments about the use of an AI system and the accountability arrangements and support learning
5. **Impartiality—'empower independence'**: AI accountability needs impartial and independent monitoring, judgement and redress

[2] AIPAS ('AI Accountability for Policing and Security') was funded by UKRI Engineering and Physical Sciences Research Council (Responsible AI (RAI) IA091 Grant Ref: EP/Y009800/1). For more details see: https://centric-research.co.uk/projects/aipas

mechanisms, without any conflict of interest in a personal, political, financial and functional way. Any conflict of interest must be identified and addressed

6. **Proof—'follow the evidence'**: a forensic approach to all aspects of AI systems and of the accountability process itself, demanding and following clear evidence; The quality of that evidence should reflect the potential impact of the AI system's use/non-use and mirror the standards of operational evidence gathering in terms of integrity, credibility and continuity
7. **Enforceability and Redress—'make it right'**: external legal and procedural routes for complaint and challenge, but also internal mechanisms for individual enforceability and redress (such as professional and policy standards) and contractual arrangements
8. **Compellability—'make it work'**: oversight bodies must be in position to make the AI accountability arrangements work giving relevant bodies the ability to compel the sharing of necessary information and evidence
9. **Explainability—'describe, demonstrate, demystify'**: provide information about AI capabilities, their purpose, and impacts in a meaningful way that is easily understood by the relevant participants/audience
10. **Constructiveness—'aim for better'**: stakeholders participate constructively with a shared aim of improvement; this may include considering different perspectives, inviting challenge and recognising how disagreement can lead to beneficial solutions
11. **Conduct—'hold yourself accountable'**: relates to adherence to professional standards, values and expected behaviours which incorporate integrity and ethics
12. **Learning—'look for the lesson'**: willingness of organisations and people to improve AI in every respect through the application of (new) knowledge and insights, including the modification and improvement of systems, structures, practices, processes, knowledge and resources, as well as the development of professional doctrine and agreed standards

The principles have connections to each other. For instance, *Explainability* is linked to *Inclusivity*, as the way *Explainability* is enacted depends on the number, type, and needs of the broad stakeholder

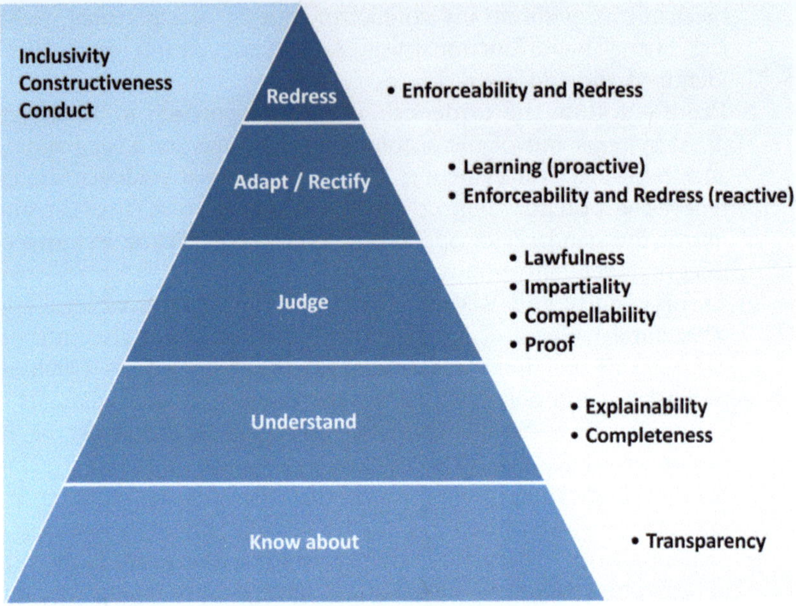


**Fig. 5.2** Relevance of the principles for the AI accountability process

landscape. In a similar way, *Learning* as well as *Redress* require *Constructiveness*, if the police aim to meaningfully engage with lessons from problematic or faulty AI deployments and with individuals/groups that have encountered negative consequences. However, the 12 principles address distinct aspects in the accountability process (Fig. 5.2), which means all principles are required to fulfil policing's AI accountability obligations.

### 5.3.2 *Operationalisation in eight themes*

To make AI accountability meaningful, it is necessary to assess how well its different principles and obligations are fulfilled. An important part of the AIPAS project was the operationalisation of the 12 principles into discrete criteria that can be judged and evaluated. We did so as a set of targeted questions that establish whether AI accountability is achieved in a specific area. The questions are formulated in a way that they can be answered as 'yes' or 'no', which allows the systematic, quantitative analysis of how far AI accountability achieved, as well as the identification of current gaps.

Across the 12 principles, the questions cover eight themes that cluster the principle-related questions according to the type of information required:

1. **AI system**: defines what the AI system does, how it functions, its purpose, limitations and who is responsible for its development and oversight
2. **Data**: focuses on the types, sources, and handling of data used in the design and deployment of the AI system
3. **Laws and regulations**: ensures that the use of AI complies with applicable laws, codes of practice, regulatory requirements and policies
4. **Risk assessment and management**: identifies risks, harms or biases associated with the AI system and implements mitigation measures
5. **Oversight and redress process**: establishes mechanisms for independent scrutiny and AI accountability, allowing stakeholders, particularly the public, to challenge decisions and seek corrections, ensuring just outcomes
6. **Stakeholders**: involves mapping and engaging all relevant parties who are involved in or affected by the AI system
7. **Accountability evidence:** focuses on documenting decisions, processes and actions taken to prove AI accountability
8. **Awareness and learning**: encourages continuous improvement through feedback, review, and learning

The benefit of the theme-based approach for end-users is that it collects AI accountability information in a more familiar 'policing' way. The translation of principles into themes and concrete questions further offers a standardised approach to AI accountability assessments, which is a foundation for the implementation of software-aided assessments solutions as presented in Chap. 9.

## 5.4 Conclusion

This chapter offers a practitioner-focused approach to define and implement accountability for AI in policing. The AI accountability concept presented here has been co-developed with policing practitioners, as well as experts in relevant fields from law to ethics and industry, ensuring that it is comprehensive and relevant for strategic, tactical and operational levels of AI deployments. This approach has been validated in Europe and the UK, as well as with subject-matter experts and citizens across more than 28 countries. These findings show that it is 'jurisdiction-neutral', i.e. works across contexts and deployment scenarios.

Our approach AI accountability is specifically aimed at the practical implementation in policing. The subsequent chapters offer concrete advice on how to implement AI accountability across specific topics (organisational practices: Chap. 6, lifecycle, stakeholder and citizen considerations: Chaps. 7, 8, 10), as well as software-supported solutions (Chap. 9).

**Acknowledgements** Work in this chapter was partially funded under the AIPAS project ('AI Accountability for Policing and Security'; UKRI Engineering and Physical Sciences Research Council, Responsible AI (RAI) IA091 Grant Ref: EP/Y009800/1).

## References

1. A. Agarwal, H. Agarwal, A seven-layer model with checklists for standardising fairness assessment throughout the AI lifecycle. AI Ethics **4**, 299–314 (2024). https://doi.org/10.1007/s43681-023-00266-9
2. B. Akhgar, P.S. Bayerl, K. Bailey, R. Dennis, S. Heyes, A. Lyle, A. Raven, F. Sampson, M. Gercke, Accountability principles for artificial intelligence (AP4AI) in the internal security domain: summary report on expert consultations (AP4AI Report 1) (CENTRIC, 2022). Retrieved from: https://ap4ai.eu/sites/default/files/2023-09/AP4AI_Report1_27Jan2022.pdf
3. B. Akhgar, P.S. Bayerl, K. Bailey, R. Dennis, H. Gibson, S. Heyes, A. Lyle, A. Raven, F. Sampson, Accountability principles for artificial intelligence (AP4AI) in the internal security domain: AP4AI framework blueprint (AP4AI Report 2) (Europol Innovation Lab, 2022). Retrieved from: https://ap4ai.eu/sites/default/files/2023-09/AP4AI_Framework_Blueprint_22Feb2022.pdf

4. B. Akhgar, P.S. Bayerl, G. Mounier, R. Linden, B. Waites, AP4AI: accountability principles for artificial intelligence in the internal security domain. Eur. Law Enforcement Res. Bull. **22**(6) (2022) https://bulletin.cepol.europa.eu/index.php/bulletin/article/view/567
5. B. Akhgar, P.S. Bayerl, How to do AI accountability—and why it's worth the effort, in *The Police Foundation. Policing and the Fourth Industrial Revolution. Conference Report of the Cumberland Lodge Police Conference* (2024), pp. 16–17
6. P.S. Bayerl, M. Obst, B. Akhgar, International Citizen Consultation on AI Accountability in Policing (AP4AI Report 3) (CENTRIC, 2023). Retrieved from: https://ap4ai.eu/sites/default/files/2023-10/AP4AI_International-Citizen-Consultation_October2023%28online-version%29.pdf
7. B.B. Watch, A Closer Look at Experian Big Data and Artificial Intelligence in Durham Police. Online Article published (6 Apr 2018) (2018) https://bigbrotherwatch.org.uk/blog/a-closer-look-at-experian-big-data-and-artificial-intelligence-in-durham-police/
8. K. Blount, Using artificial intelligence to prevent crime: implications for due process and criminal justice. AI Soc. **39**, 359–368 (2024). https://doi.org/10.1007/s00146-022-01513-z
9. F. Cech, The agency of the forum: mechanisms for algorithmic accountability through the lens of agency. J. Respons. Technol. **7–8**(2021), 100015 (2021). https://doi.org/10.1016/j.jrt.2021.100015
10. R.A. Duff, Moral and Criminal Responsibility: Answering and Refusing to Answer (2017). https://ssrn.com/abstract=3087771
11. Fair Trial, FOI reveals over 12,000 people profiled by flawed Durham police predictive AI tool. Online Article published 15 Aug 2024 (2022). https://www.fairtrials.org/articles/news/foi-reveals-over-12000-people-profiled-by-flawed-durham-police-predictive-ai-tool
12. M. Gerlich, AI tools in society: impacts on cognitive offloading and the future of critical thinking. Societies **15**(1), 6 (2025). https://doi.org/10.3390/soc15010006
13. High-Level Expert Group on Artificial Intelligence (AI HLEG), The Assessment List for Trustworthy Artificial Intelligence (ALTAI) for self-assessment (2020). https://digital-strategy.ec.europa.eu/en/library/assessment-list-trustworthy-artificial-intelligence-altai-self-assessment
14. G. Lennon, N.R. Fyfe, J. McNeill, F. Sampson, *Principles for Accountable Policing* (The Police Foundation, 2023) https://www.police-foundation.org.uk/wp-content/uploads/2010/10/accountable_policing_FINAL.pdf
15. A. Schedler, Conceptualizing accountability, in *The Self-restraining State: Power and Accountability in New Democracies*, ed. by A. Schedler et al. (1999), pp. 13–28

**Petra Saskia Bayerl** is Professor of Digital Communication and Security and Head of Research at CENTRIC. Her research addresses societal implications and governance of new technologies in security and policing. She has been co-lead for both the AP4AI and AIPAS projects and lead for national and EU-projects to improve the protection against online harms.

**Babak Akhgar** is Professor of Informatics, Director of CENTRIC and a Fellow of the British Computer Society. He has extensive and hands-on experience in the development, management and execution of KM projects and large international security initiatives (e.g. combatting terrorism and organised crime, AI in Policing, cyber security, public order and cross-cultural ideology polarisation). He has more than 100 refereed publications in international journals and conferences on information systems, specifically focusing on knowledge management (KM). Babak was awarded an OBE in June 2022, the second highest order of the British Empire award, in recognition of his services to security research.

PART II

# Practical Considerations

CHAPTER 6

# Organisational Best Practices for AI Accountability and Acceptance

*Christopher Rowley and Martin Snowden*

## 6.1 Introduction

Artificial Intelligence (AI) in the general UK workplace began to gain impetus from 2010 onwards; however, it is only in recent years that it has begun to gain a foothold in the different pillars of policing and law enforcement. This 'delay' may have been through fear amongst senior police leaders of the unknown and also concerns following public backlashes to early experimentation with predictive and surveillance tools without any obvious accountability mechanisms.[1] Further, financial restraints, which have been a mainstay in policing since 2010, often prevented and still limit technological investments.

[1] Examples include Predictive policing tools, and the use of Automatic Facial Recognition tools at Notting Hill Carnival, e.g. https://www.bbc.co.uk/news/articles/cyvn5e93z8jo.

C. Rowley (✉) · M. Snowden
CENTRIC (Centre of Excellence in Terrorism, Resilience, Intelligence and Organised Crime Research), Sheffield, UK
e-mail: christopher.rowley@shu.ac.uk

 
B. Akhgar and P. S. Bayerl (eds.), *Accountability for Artificial Intelligence in Policing and Security*, Security Informatics and Law Enforcement, https://doi.org/10.1007/978-3-032-41284-3_6

There is now an inevitability that AI will become the norm in policing, whether it is used to predict the location of the next crime, identify a potential terrorist or to run more efficient and cost-effective back-office functions. Identifying, purchasing and implementing the most effective AI systems will ensure that police forces can better offer the level of protection society demands (see Chap. 10), whilst delivering value for money, against ever decreasing public finances.

AI in policing has developed a negative connotation within sections of the public and media, emanating primarily from its use of AI processes in Live Facial Recognition operations at large-scale public events. Associated court rulings in judicial processes criticising, and indeed in some cases finding unlawful police practices pertaining to the retention of biometrics and facial images,[2] added to the suggestion and labelling that the use of AI was 'state intrusion' on an individual's right to a private life (e.g. [3]). There are also concerns regarding the quality of the data used by AI systems, which may cause the system to demonstrate an inherent bias towards certain groups in society, thereby adversely affecting the way they are policed (e.g. [2]). This societal fear is grounded in evidence-based research and, if not eradicated during the development stage of the AI system, could lead to a loss of trust and confidence in policing.

This is significant, as trust in policing, and the wider criminal justice system is already low and indeed declining.[3] Trust and confidence have been eroded over many years, for a wide variety of reasons, influenced by both national reporting and media coverage or personal experiences. Examples cited include the lack of visible policing, poor service delivery, rising crime, poor detection rates, significant backlogs in the judicial system, as well as racism, sexism and criminality committed by serving police officers [1].

Restoring public trust and confidence as a whole is a complex issue and challenging task for policing. However, any steps, however small or

[2] E.g. Courts and Judiciary Tribunal report—The Queen (on the application of Edward Bridges) (Appellant) v The Chief Constable of South Wales Police (Respondent) & others [2020] EWCA Civ 1058 On appeal from: [2019] EWHC 2341 (Admin) Judges: Sir Terence Etherton MR, Dame Victoria Sharp PQBD, Lord Justice Singh.

[3] College of Policing (online) Trends Report 22nd April 2025 (with highlights from the Crime Survey)—Trust in policing has been in steady and constant decline since 2017/18—people who agreed that the police in the local area do a 'good job' or an 'excellent job' fallen 13% points from 62% to 49%. www.college.police.uk/guidance/improving-public-confidence-police/trends.

incremental to improve operational performance and release officers and staff back to the 'front-line' must be considered.

With the ever and rapidly evolving potential of AI systems, police forces across the United Kingdom, indeed globally, will therefore be considering whether AI processes can enhance their front-line service delivery, to prevent and detect crime for example, whilst still building trust and confidence in the communities they serve. They may also be fearful of the unknown: Can the AI system be managed, will it become too powerful and take over? These questions are not only asked by police chiefs but by leading politicians and academics. Yet an alternative way that they could consider AI is, 'can I run a highly effective, efficient and legitimate organisation without it'?

## 6.2 Strategic Approach Combining Short-, Medium- and Long-Term Focus

AI systems are being used to great effect in some forces across the UK and other countries. Several UK forces have adopted AI to deal with low-level enquiries which would normally be answered either by front office staff or via the 101 system.[4] It is also being used to predict the location of the next crime such as burglary offences, allowing police forces to allocate valuable policing resources to locations with the highest likelihood of repeat offences taking place. AI can be used to analyse large volumes of digital forensic data, which will show links between offenders and offences and has the ability to identify known wanted offenders during live events, allowing officers to arrest the offender and to prevent the possibility of re-offending (see Chap. 12). It can also revolutionise the back-office functions provided to policing: AI can increase productivity, accuracy and timeliness, in areas such as vetting, firearms licencing, finance, recruiting, human resources and all other support services which enable police forces to operate.

Of course, the purchase and deployment of an AI system must be carefully considered to achieve the operational aims whilst still maintaining, indeed potentially restoring, public trust.

[4] NPCC News Report 17th March 2026—West Yorkshire Police, in partnership with the NPCC Digital Public Contact (DPC) programme, launching a new artificial intelligence capability designed to extract insight from historic control room calls. AI to unlock intelligence from control room calls.

To support the responsible procurement/development and subsequent deployment of the most appropriate AI system, organisations should develop a short-, medium- and long-term strategic intention based around the 12 AIPAS Accountability principles and the eight themes as outlined in Chap. 9. The 12 AIPAS principles (Lawfulness, Proof, Completeness, Inclusivity, Explainability, Transparency, Compellability, Enforcement and Redress, Impartiality, Constructiveness, Conduct and Learning) and the eight practical themes (AI system, data, laws and regulations, risk assessment and management, oversight and redress, identification and management of stakeholders, AI accountability evidence, and adaptability and learning) will guide the organisation and the AI user. Specifically, they will assist in the procurement of AI systems, their implementation and use, whilst guiding the organisation in the development of transparent and accountable processes. The adoption of these principles and themes will allow the organisation to develop policy and best practices which will help them to carefully consider their data source and remove bias and may improve the decisions and actions by police officers and staff.

Ensuring that human oversight is a key part of the final decision-making process will be a critical factor to build trust and confidence, when the AI system makes a recommendation relating directly to any human interaction. This process will also support the development of transparent governance processes, which can be shared with and scrutinised by communities. This level of transparency and consultation will assist in increasing both trust and confidence in the use of AI and policing more broadly. Conversely, failure to implement essential processes around the provision of information, public engagement and effective and timely remedies may have a negative impact on such trust and confidence.

The procuring body and party to any commercial contract will be the elected local policing body (e.g. regional Mayor or Police and Crime Commissioner in England and Wales; ref- practical-accountability-chapter), unless they grant the force's chief officer express authority to enter into procurement contracts. This is an important consideration, both in terms of the procurement parameters (choice of supplier, etc.) and also the decision to use functional AI within their policing area.

However, overall legal responsibility and thus accountability for the organisational AI strategy sits with the Chief Operating Group and the chief officer. The strategy should be developed by a Strategic Governance Group (see Fig. 6.1), which is chaired by a senior responsible officer.

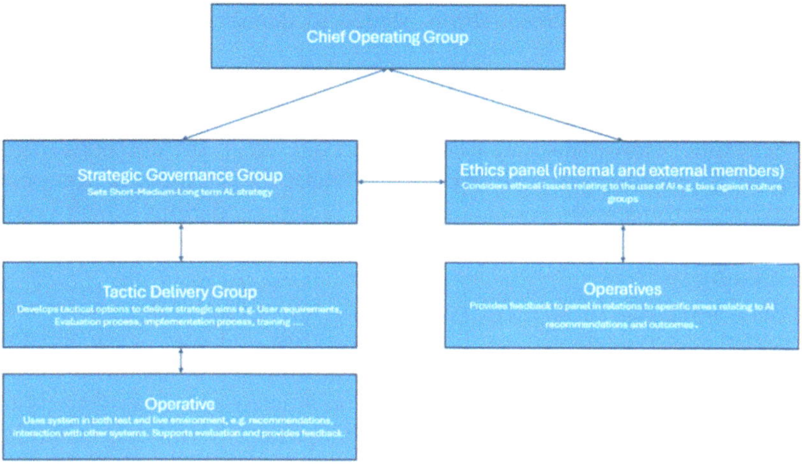


**Fig. 6.1** Illustration of best practice for force governance and delivery model

Good governance would suggest that there should be both a Strategic Governance Group and Tactical Delivery Group. As their titles suggest, the strategic group will develop the strategy, informed by the tactical group, which will be formed by subject matter experts and operational users. The tactical group will develop work packages to deliver the strategy. This governance structure will then develop key milestones, which will ensure that the short-, medium- and long-term strategy are delivered on time and in accordance with the overall AI strategy.

### *6.2.1 Elements of Short-Term AI Strategy*

The short-term strategy should consider which areas of the organisation will benefit from an AI system and how that will in turn benefit the community being policed. Considerations may be areas which are low risk, with no immediate potential for adverse decision-making (e.g. a back-office function). Conversely the organisation may have an issue with significant risk, which will also benefit from AI deployments (e.g. large volumes of digital data linked to multiple organised crime group investigations). The strategy should also include considerations regarding the most appropriate governance structures (as detailed above), and the internal and external consultation needed to demonstrate openness and transparency.

Internal and external stakeholder consultation can add a more diverse perspective, which may not be obtained by only consulting internally, where some individuals may have a vested interest in the decisions made regarding the adoption of an AI system. Whilst the organisation does not need to automatically adopt any of the recommendations which come from that consultation, they should carefully consider the implication of not doing so.

The adoption of any AI system with the support of internal and external stakeholders can assist in building trust and confidence and the future adoption of more risk-based AI systems. It will also help future relationships where an AI recommendation is acted upon and there is an adverse wider public reaction. The early adoption of multi-cultural ethics panels, which has both internal and external representatives (experts and lay persons), may also enhance the future adoption and use of an AI system. The panel should have clear terms of reference, and members should be encouraged to be professionally curious and to challenge decisions. By doing so they will add real value and ensure the organisation continues to evolve in its thinking, actions and policy development.

### 6.2.2 *Elements of Medium-Term AI Strategy*

The medium-term strategy should learn from AI systems adopted by the organisation and in other organisations. The organisation should look up and out across both public and private sectors to learn their lessons and, where appropriate, develop partnerships to develop and deliver efficiencies and improve products and policies. This will ensure that police organisations do not become complacent and continue to adopt the best products on the market, which delivers to the needs of their communities and their people.

### 6.2.3 *Elements of Long-Term AI Strategy*

The long-term strategy should continue to evolve, ensuring it reviews the lessons learned from both the short- and medium-term strategies against both the deployment and procurement assumptions and assurances. Where there is clear evidence that an AI system delivers an enhanced effective, efficient and legitimate service which prevents and detects crime in an ethical, open and transparent manner, where it eliminates bias and supports the building of trust and confidence between policing and those it serves, the organisation should gain confidence from this and consider further areas of policing, which had previously not been considered as suitable for AI deployments.

## 6.3 General Considerations for Continued AI Accountability before and during Deployment

When considering which AI system to purchase or develop, the organisation should have very detailed and clear user requirements, which are likely to be aligned to force strategic priorities. For example, if the system is going to be used to identify issues relating to possible racial profiling during stop and search, how will this be highlighted, what are the markers that will be used, how will this AI system report this, and to what level of accuracy? Moreover, if an AI system is capable of racial profiling, how will the risks of reverse use be addressed and communicated to communities? Another example is predictive policing and predicting future offending. Prior to the adoption of any AI system, the organisation will need to ensure that data has been cleansed to allow for a high level of certainty. It will also need to consider any potential bias which may be in its crime and intelligence systems prior to considering recommendations relating to future offending. Whilst the cleansing of data was once seen as an arduous task that is both expensive and time-consuming, this is no longer the case; there are AI solutions, which organisations are currently using, to cleanse their data in hours or days rather than weeks or months.

After all ethical issues have been considered, detailed user requirements have been developed, and a product has been identified, the organisation should rigorously test the AI system using historical or live data. A detailed evaluation model should be developed using the eight AIPAS themes (see Chap. 9) to ensure the AI system delivers the desired outcomes for the organisation. This will allow an experienced operative to check each outcome in a test environment against an agreed evaluation process. Whilst this may elongate the adoption of the system, it will also allow the organisation and the ethics committee to understand and, where appropriate, challenge the outcomes.

Examples, where the AI system may not work as initially believed, include demonstrating biases, extracting data from unknown sources, or giving unrelated or rogue recommendations. Conversely, the outcomes may offer improved results in areas which were not expected, such as the early identification of vulnerability or poor decision-making by operatives which have become custom and practice within the organisation and require additional training.

The findings from the test environment evaluation should be reported into the governance structure and the ethics committee prior to the AI system being adopted into a live environment. When the AI system is adopted into live operational or business areas, consideration should be given to adopting it in incremental stages (e.g. during particular times of the day, so it can be monitored in real time or in areas with lower demand), again allowing the operative to monitor and check all recommendations without the addition of high demand (e.g. a call handling environment).

After the AI system has been launched in a live environment, the evaluation of the AI system should continue with regular dip sampling by supervisors to ensure the AI system is functioning as expected and that the operative is adhering to policy, which should include ratification of AI recommendations where the decisions relate to human interaction and their individual needs. Operatives of any AI system should be encouraged to provide regular feedback to the strategic and tactical governance groups; they should also be invited to speak with the ethics committee as part of the evaluation process. If no feedback is forthcoming, then it should be sought by the groups to ensure any early lessons can be learnt or issues can be resolved to improve the AI system, its recommendations and the service to communities.

## 6.4 Managing Internal and External Concerns

The adoption of any new AI system may bring with it, for some, an element of fear that it will result in a reduction of staff within the organisation. This fear will likely become a reality in many public and private sectors in the coming years; however, this cannot be a bar to prevent the development and use of AI systems to improve the service provided to victims and communities. To assist with staff engagement in this regard, and also wider public awareness, the organisation should develop an internal and external communication strategy.

Internally, the communication strategy will help to ensure employees, officers, unions and staff associations are informed of the developments and why they are necessary to improve the service to the public through innovation, particularly when public sector funding is being cut. For some this will not reduce their anxieties or fears, for others it will present a new opportunity to retrain in a new role within their own organisation or in another organisation. Externally, the communication strategy will help to inform the community of where the AI system is being used, how it is improving the service the organisation provides to them and detail key principles and requirements such as complaints management and ongoing reviews.

The key to the success of the AI system delivering the outcomes expected is the extent to which it meets the requirements of the 12 AI Accountability principles as put forward in AIPAS: not only at procurement and implementation stages, but also the training of all operatives. The tactical group should develop a high-quality training programme, which not only trains the operative in how to use the system but also teaches the ethical considerations that must be checked when a recommendation is made (i.e. bias against a particular group, is the recommendation lawful, is it ethical?). Good quality training will give the operatives the confidence they need to use the system and to see where the AI system may not be operating as expected. This will enhance the quality and speed of the decisions/outcomes and improve the service provided, whilst building confidence and trust.

As this is a long-term strategy, with short- and medium-term strategic goals, the governance structure and evaluation process will become a permanent element in the overall organisational structure, reporting when agreed to the Chief Operating Group.

## 6.5 Conclusions

It will take some time for both the public and the organisation to feel comfortable with the adoption of AI systems. To develop the trust and confidence needed, new AI systems must be carefully procured/developed through an open, transparent and accountable governance structure, based on considerations within the AIPAS 12 AI accountability principles and eight themes. This chapter offered practical recommendations for police leaders and police forces aimed to support the planning and preparation of the organisation, as well as the proactive risk management of AI implementations. (More detailed, operational guidance can be found in the Implementation Guidance for AI in Policing in the Appendix).

The core goal is that an effective AI accountability regime will be integrated into the wider organisational audit arrangements (internal and external) and the relevant inspection regime. This will allow the organisation to incrementally implement and robustly evaluate new AI systems, which can support an increase in trust and confidence in the organisation whilst improving overall organisational effectiveness, efficiency and legitimacy.

## References


1. P. Hanway, O. Hambly, Public perceptions of policing: a review of research and literature. Home Office Research and Analysis Report (2023). https://www.gov.uk/government/publications/public-perceptions-of-policing-a-review-of-research-and-literature/public-perceptions-of-policing-a-review-of-research-and-literature
2. A. Shapiro, Predictive policing for reform? Indeterminacy and intervention in big data policing. Surveil. Soc. **17**(3/4), 456–472 (2019) https://ojs.library.queensu.ca/index.php/surveillance-and-society/index
3. S. Srivastava, S. Shukla, K. Rohira, The ethics of predictive policing. Analyzing privacy concerns and the social systemic bias, in *Emerging Technology and the Law: Bridging the Gap Between Innovation and Regulation*, ed. by P. Pokhariyal, A. Patel, N. Singh Nautiyal, A. Kumar Mishra, 1st edn. (Routledge, 2026). https://doi.org/10.4324/9781003634591



**Christopher Rowley** retired as Lancashire Police's Chief Constable following a 31-year career in policing. A graduate of Cambridge University, he is passionate about identifying new and innovative ways of preventing people becoming victims by adopting evidence-based academic evidence. He currently is Strategic Lead for Policing in CENTRIC (Centre of Excellence in Terrorism, Resilience, Intelligence and Organised Crime Research).

**Martin Snowden** is Senior Research Fellow, and Counter Terrorism Lead, at CENTRIC. He is a former Detective Chief Superintendent and was Head of Counter Terrorism Policing North-East between 2017 and 2022, responsible for delivering all elements of the Government's Contest strategy and CT policing across the North East region of England. His CENTRIC is focused on current policing challenges, AI accountability and the reduction of online harms.

CHAPTER 7

# AI Accountability Themes and Lifecycle: Practical Guidance

*Christopher Spencer and Petra Saskia Bayerl*

## 7.1 Introduction

Lennie Hoffman in *DPP v McKeown* notoriously said that 'one needs no expertise in electronics to be able to know whether a computer is working properly' (as cited in [4], p. 65). Of course, computer scientists would disagree, and one would argue so would a typical user in 2026, particularly considering the UK Horizon scandal, where faulty accounting software used by the UK Post Office led to hundreds of wrongful prosecutions. The Horizon scandal exemplifies accountability failings, not just in the technology that was used but also across the entire ecosystem. As Chalmers [2] points out, how do we ensure that accountability 'is "built in" to the opportunities and complexities that new technology like AI will bring?' The capabilities of computers have changed and will continue to change, and with it so will the environments they shape and that they are shaped by [3].

C. Spencer (✉) · P. S. Bayerl
CENTRIC, Sheffield Hallam University, Sheffield, UK
e-mail: christoper.spencer@shu.ac.uk



B. Akhgar and P. S. Bayerl (eds.), *Accountability for Artificial Intelligence in Policing and Security*, Security Informatics and Law Enforcement, https://doi.org/10.1007/978-3-032-41284-3_7

Before modern AI systems, computers relied on predefined rules and static programming and could only execute tasks that were explicitly coded by humans. With AI, however, systems can now learn from data, adapt to new information and improve their performance. This shift enables AI to recognise language, identify images and make predictions. In a police and security context, these benefits come with added accountability demands. As highlighted in Chaps. 2 and 3, policing has a plethora of actors who they are accountable to and for, and whose accountability requirements differ. For policing, this means accountability needs to be expansive, robust and iterative.

Although this chapter is written from a UK policing context, its purpose is not to present the UK position as exceptional or complete; rather, the UK context is used as an applied case as it provides a mature policing environment and one where accountability is central to all aspects of operation (for details see Chap. 2). The underlying accountability logic is transferable to other jurisdictions, and indeed universal in application, on the condition that aspects like legal bases, regulatory duties, oversight arrangements and community expectations are incorporated into native contexts. The central argument of this chapter is that AI accountability in policing must be embedded across the AI lifecycle and evidenced through practical accountability themes. The chapter refers back to the AI lifecycle through the prism of pre-deployment, deployment, and post-deployment which is the practical sequence through which accountability commitments begin (see Chap. 5). It then details the eight practical accountability themes formulated in AIPAS that translate high-level accountability principles into daily policing practice. For completeness, the chapter should be read in conjunction with the guidance for implementing AI accountability in policing ([1]; provided in the Appendix in this book) and the conceptual framing in the starting chapters of this book of AI accountability (Chaps. 2, 3, 4) and AIPAS (Chap. 5).

## 7.2 Why Is Accountability Important Across the AI Lifecycle?

The AI lifecycle describes the various stages that an AI system experiences from initial conception through deployment, operation and eventual decommissioning. Interpretations of this lifecycle vary significantly. For example, De Silva and Alahakoon [6] conceptualise the lifecycle in three primary phases: design, development and deployment, subsequently detailing 19 subordinate stages. Ng et al. [12] detailed a data-centric approach specifically designed for healthcare AI systems, proposing five key stages: data creation, data acquisition, model development, evaluation and deployment. Additionally, the OECD ([13], p. 13) offers another perspective, defining the AI lifecycle through four phases of design, data and models, verification and validation, deployment and operation and monitoring. ISO/IEC [9] arguably offers the most complete example, being planning, development, deployment, operation, maintenance and disposal.

Despite these varied interpretations, each approach divides the lifecycle into stages where responsibilities can be inferred. We argue that for each stage specific lines of accountability exist. Such clarity is essential, particularly in a policing context, where public trust and confidence directly influence operational effectiveness [16]. Indeed, the legitimacy of UK police, guided by the Peelian Principles, relies fundamentally on public support, and failures in accountability can greatly undermine legitimacy.

Accountability across the AI lifecycle is not merely beneficial but essential, as it ensures that the technology and those who use it remain transparent, ethically aligned, trustworthy and serve the public lawfully. To help navigate policing through AI lifecycle accountability considerations, we have devised the lifecycle into three stages of (1) *pre-deployment*, which can include aspects like development and procurement, (2) *deployment*, which is the operationalising of the AI system and can include migration of an AI system, and finally (3) *post-deployment*, which involves the eventual end state of the AI system being decommissioned (see also Chap. 5). Table 7.1 demonstrates the AI lifecycle of pre-deployment, deployment and post-deployment and compares against other interpretations. The purpose here is not to rank which interpretation is correct but rather indicate the similarities when segmenting the AI lifecycle along thematic areas.

**Table 7.1** Different interpretations of the AI lifecycle

| | *Pre-deployment* | *Deployment* | *Post-deployment* |
|---|---|---|---|
| Bayerl and Akhgar (2026, see Chap. 5) | – Design<br>– Procurement | – Deployment<br>– Modification | – Migration<br>– Decommissioning |
| De Silva and Alahakoon [6] | – Design<br>– Development | – Deployment | |
| Ng et al. [12] | – Data creation<br>– Data acquisition<br>– Model development<br>– Evaluation | – Deployment | |
| OECD [13] | – Design data and modelling<br>– Verification and validation | – Deployment<br>– Operation<br>– Monitoring | |
| ISO/IEC [9] | – Planning<br>– Development | – Deployment<br>– Operation<br>– Maintenance | – Disposal |

## 7.3 Considerations across AI Lifecycle Stages

### 7.3.1 *Pre-deployment*

AI accountability begins before an AI system is ever put into use, namely with procurement or design. Decisions made at this stage lay the foundation for how responsibly and effectively AI will serve policing and the public. Accountability during pre-deployment requires foresight across the entire AI lifecycle, and by taking an accountability-focused approach policing can reduce downstream impacts across policing, including unnecessary costs in both time and money.

A pragmatic approach should ***answer fundamental questions*** like:

- What are the accountability requirements put forward by law or relevant policies impacting on decisions to procure or design and ultimately use an AI-driven approach?
- What are the needs and requirements of policing and their stakeholders, especially the public, that justify the procurement or design of an AI capability?

From this position, a list of requirements can be established that can provide a baseline for the accountability-aligned capabilities an AI system must demonstrate. The relationship between accountability requirements and technical requirements should therefore be explicit. If the accountability requirement is explainability, the technical requirement may include meaningful output descriptions.

However, AI accountability cannot be limited to the AI system alone; it must also extend to the human decisions that govern its use and the processes and procedures for its use. Alongside technical requirements, non-technical considerations of AI accountability must thus be embedded in all procurement or design decisions. Frameworks such as Data Protection Impact Assessments (DPIAs) and existing policy decisions already provide useful mechanisms for this and—in the absence of comprehensive AI legislation in the UK—currently represent a 'best fit' solution (for a discussion of regulations in the EU AI Act see Chap. 4). However, policing cannot rely solely on laws. Instead, policing must include foresight by embedding forward-looking accountability structures that anticipate evolving risks and expectations surrounding AI. For this reason, a *federated approach towards pre-deployment* should be adopted, and clear lines of responsibility, leadership and involvement should include stakeholders responsible across a range of areas, as demonstrated in Table 7.2.

Table 7.2 demonstrates high-level considerations for policing that should be answered in order to inform a strategy on how policing will approach the adoption of an AI system. Each stakeholder will have specific, sometimes unique accountability requirements, and these requirements must be indicated prior to procurement and development to ensure accountability needs that are met across the AI lifecycle. Simply put, it is important to ensure that the correct people are involved and from early on. By taking this approach, policing will have a solid understanding prior to entering procurement or design discussions, and teams within policing will be aware of the high-level information needed to develop an AI procurement strategy.

#### *7.3.1.1 Procurement-Specific Considerations*

An AI system can be developed internally, through a third-party (i.e. externally), or a hybrid of the two (co-created). AI systems developed by a third-party will necessitate careful communication and consistent engagement to ensure that the AI system meets the needs of policing

**Table 7.2** Considerations across AI lifecycle stages

| *Area* | *High-level aspects that should be considered* |
|---|---|
| *Pre-deployment* | |
| Procurement | Supplier background checks that examine aspects such as transparency, security, lifecycle support, cost, update arrangements, audit access and exit provisions |
| System design | AI systems functionality limitations, testing approach, human oversight and how accountability evidence will be recorded |
| Process design (e.g. training, redress procedures) | Specify the operational workflow around the system, including training, authorisation, escalation, redress, decision recording and supervision |
| *Deployment* | |
| Deployment and operation | Define pilot or sandbox arrangements, user roles, performance monitoring, operational thresholds, reporting routes and procedures for pausing or restricting use |
| Legal and compliance | Evidence of legal basis, human rights impact, data protection needs, equality consideration, procurement compliance and force policy alignment |
| Audit, inspection and wider stakeholder requirements | Understand stakeholders' requirements and how/whether accountability functions of the AI system support audit and inspection requirements |
| Data and data protection | Understand information on data provenance, lawful use, quality, representativeness, retention, storage, access, sharing, deletion and bias testing |
| Support and maintenance | Identify who maintains the system, what maintenance is included/needed and how updates are approved |

(continued)

**Table 7.2** (continued)

| *Area* | *High-level aspects that should be considered* |
|---|---|
| *Post-deployment* | |
| Migration | A plan for transferring data and systems without losing accountability evidence or operational tempo |
| Decommissioning | Arrangements for decommissioning and clauses on archiving, deletion, supplier-held data, evidential preservation and methods to review data on decommissioned AI systems |

and AI accountability requirements. This is pivotally important; one does not need to look far to see procurement blunders due to miscommunication between users, stakeholders, suppliers and buyers. For example, FiReControl was a UK Emergency Service project to replace 46 Fire and Rescue control rooms with nine purpose-built regional control centres and supporting IT systems. An inquiry found that the project failed because of underestimations in cost and complexity, leadership and management issues and lack of support from stakeholders—particularly its users (National Audit Office, 2011a). Similar issues could be witnessed in the procurement and failed deployment of the London Ambulances Service's Computer Aided Dispatch project where an inquiry found that the project exposed weaknesses in procurement, stakeholder engagement (including user consultations), training, operational readiness and technical resilience [17].

In the UK, *procurement rules and obligations* are set out in laws, regulations and policies [5, 14]. Similar directives can be observed elsewhere such as the European Union, which separate procurements by security and defence [7, 8]. Agnostic to their originator, the crux of these laws and regulations is to ensure that fair and transparent procurement takes place within public sector organisations, and that the public get value for money. For policing, this means that when an AI system is procured, it must be able to demonstrate that the system was selected through a fair, transparent, lawful and evidence-based process. It also means policing must safeguard themselves and by extension public money, with correct contract arrangements and tested supplier claims.

Outside of any procurement process governed by laws and regulations and purely from a policing perspective, the first step in accountable procurement is to clearly define the need, purpose and necessity of the AI tool or capability. Policing should be able to articulate the specific problem or operational gap that the AI is intended to address and justify why an AI solution is appropriate. This typically occurs through a business case or requirement documents. Establishing a clear purpose is crucial for several reasons: it ensures that the technology is truly needed and proportionate, guides the selection for the right tool features and sets boundaries on how the tool will be deployed. In practical terms, policing should be able to answer, 'Why do we need this AI system and what outcomes do we expect (if used or not)?'

This *demonstration of necessity and proportionality* is not just good practice but required to comply with human rights obligations and data protection principles. A statement of purpose also supports transparency as it provides a basis to explain to stakeholders what the tool is meant to do and not do. Once policing has this understanding evidenced, it can then enter the procurement process which governs it.

#### *7.3.1.2 Design and Development-Specific Considerations*

Typically, when an AI system is internally developed, policing will have more control and greater transparency over the design and functionality. Yet internally developed systems still need to be held to the same levels of accountability and scrutiny as third-party systems, which should include expert reviews from legal and technical professionals. The sooner policing can communicate their requirements to the correct stakeholders (and vice versa), the smoother the development process will be. Table 7.3 offers a summary of the pragmatic considerations when designing an accountability-aligned AI system for policing. It presents accountability features that an AI system should possess to ensure that policing has a clear understanding of how the AI system is configured, its functions, inputs and outputs and impacts.

### *7.3.2 Deployment*

While procurement can be coined as a 'planning' stage, deployment can be referred to as the 'doing' phase and argued to be the most challenging phase in the lifecycle [10]. Along the AI lifecycle, there will be a period where pre-deployment meets deployment, and either the AI

**Table 7.3** AI accountability considerations for the AI system

| *AI system-level accountability evidence* | *What the AI system should be able to evidence* |
|---|---|
| Output explanation | How the AI system reached a particular output, akin to a methodology |
| Input explanation | What information has been entered, when it was entered, who entered it and any gaps in the data that would influence the output |
| Data provenance | Where the data came from, including the source system, dataset, record, database, sensor or user input that contributed to the output |
| Model or system version/settings | Which model, software version, configuration, ruleset, threshold or algorithmic setting was active when the output was generated |
| Decision pathway | Confidence levels attached to an output, identify uncertainty and flag where the result is close to a decision threshold |
| Data bias and limitations | Notifications if an output is affected by known data limitations such as bias or gaps |
| Human interaction | Record of user logs that can demonstrate the user of the AI system and what the output was used for. This can range from stop and search in the case of live facial recognition or creating an intelligence report with the use of crime mapping |
| Performance and accuracy | Performance information, such as accuracy, false positives and false negatives. Evidence should also display when a change has taken place, i.e. to the model, data, thresholds, rules or configuration |
| Exportable accountability record | Exportable audit record for review that includes all the points in the right-hand column |

Recognising the fact that not all who serve in policing can understand computer science or algorithmic equations, at the design level, accessibility of accountability evidence is itself a design requirement and the AI system should be developed so that its outputs can be explained at different levels (i.e. by computer scientists, the users, a jury)

system will be piloted (in synthetic or operational contexts), or when the AI system is transitioning into operation. During the pilot and transition periods, the AI system should be compared against pre-existing technologies or processes to ensure that the AI system is of benefit and working as intended. Comparison with existing processes should include measurable and qualitative criteria; for example, if the AI system was used to triage domestic abuse reports, 'how does it compare to current technologies or processes; what are the friction points; is it working as intended; what are the impacts/consequences (negative and positive, intended and unintended)?'

Comparison points could include accuracy over existing systems, volume of false positive and false negative, time to complete tasks or impact on workloads. If the AI system is filling a capability gap in policing, the efficiency of the AI system will still need to be measured (e.g. has there been an improvement since using the AI system?).

AI accountability during this part of the lifecycle, and across the entire deployment, demands continuous monitoring, feedback and where necessary adjustments to improve systems and processes. The first deployment of an AI system will create new challenges and opportunities for policing. While rigorous planning during pre-deployment and a staged and managed transition might foresee these challenges and opportunities (e.g. through risk assessments, contractual agreements and training provisions, using themes considerations described below), only the deployment will establish the completeness of this forward planning. Oversight and redress therefore are an essential aspect of any deployment, allowing reaction to a fuller understanding of the AI system's use in practice. For oversight and redress to work, policing must have processes in place where information can be passed to the intended stakeholder(s). Common ways, in which policing already achieve this, are through performance reviews and capability meetings (details and examples see Chap. 2).

For policing to be effective, and by extension the use of an AI system, consideration must be given to the public and affected communities. AI systems that are operational can have varying degrees of intrusion into public life, be it from facial recognition technologies, investigative tools or predictive crime mapping. As demonstrated in Chap. 10, acceptance of and trust in police use of AI can vary depending on a person's experiences and demographics. It also demonstrates that each deployment or use context may elicit different responses from the public. Careful management and communication can help shape the narrative around an AI deployment and more broadly around AI in policing.

However, careful planning and communication alone will not solve all potential issues in which AI may bring to policing and public AI acceptance. Deployment must also be linked to organisational learning and redress. If a system produces a wrong or unfair outcome, there must be a mechanism to identify it, report it, review it, correct it and learn from it. An important element is redress, as part of the AIPAS principles. Redress is not solely a complaints process but rather an accountability mechanism that shows that policing (or any other organisation for that matter) accepts responsibility for the consequences of its actions. Redress also provides valuable opportunities for organisational learning where patterns in complaints or a high volume of similar user concerns may indicate deeper issues. This aspect also addresses the fact that deployment decisions must remain reversible. If the AI system behaves unexpectedly, the legal context changes or if public harm outweighs operational benefit, policing must be able to pause, limit, alter or stop use. Deployment is not the end of accountability; on the contrary, it is the point at which accountability is most visible, and mismanagement can have most impact.

### 7.3.3 *Post-deployment*

Considerations for post-deployment address both migration and decommissioning situations. Some considerations will be relevant for both although with specific aspects.

#### *7.3.3.1 Migration*

Migration is often treated as a technical or administrative point in AI accountability. However, it is a point in the AI lifecycle where police must ensure that the AI system and its data, records, etc., are transferred without creating new risks (such as legal, operational or evidential). Migration in the AIPAS lifecycle refers to relocating or transferring an AI system, its data and outputs. This may involve moving the system to new infrastructure, integrating it with another platform, updating the model or changing the deployment context.

Policing should treat migration as AI lifecycle stages requiring:

- Clear decision-making records.
- Transition planning.
- Data and outcome archiving.
- Documentation archiving.
- Maintenance of access rights.

Independent of the type of migration, the move should be tested and validated in a similar manner to when a new system is being deployed. At minimum, policing should test and record whether outputs, access rights and system features remain valid, and whether performance has changed (increased or decreased). With this information, policing will be able to formulate a response, if required.

Migration can create specific risks around data retention, deletion, transfer, archiving and access. Therefore, policing must:

- Know what data exists.
- Where it is stored.
- Who controls it.
- Who has access to it.
- What must be retained.
- What must be deleted.
- What third parties must do.
- How to ensure that data remain accessible long-term.

Where suppliers are involved, contractual obligations should require them to return, delete, transfer or preserve data in accordance with agreed legal and operational requirements. Furthermore, if an AI system's' output has influenced policing decisions, stakeholders may later ask policing to explain how it worked, what data it used, what output it produced, who relied on it, what human decision followed and so on. For this reason, migration must preserve previous data that the AI system relied upon, and policing must maintain the ability to account for its previous use.

#### *7.3.3.2 Decommissioning*

Decommissioning is the stage at which an AI system is formally withdrawn from operation; there are two main contexts in which this can occur. An AI system may be withdrawn as part of a planned decommissioning strategy or unplanned if the system is not fit for purpose (e.g. produces biased, unlawful or problematic outputs). Whichever the context, policing must make best efforts to ensure that the system is safely removed without losing accountability evidence, disrupting operational activity or creating legal risk. Even when an AI system is turned off, the effects may remain as outputs in case files, intelligence assessments or operational decisions. If an AI system that has later been found to not be fit for purpose during deployment, stakeholders may later ask to review the AI system or its decisions. Therefore, during decommissioning, policing must preserve AI system data and associated accountability documentation. This can include access logs, model information, system settings, training records, risk assessments, legal assessments, performance evidence and human decision records.

Similarly to migration, the plan for managing decommissioning should also be included into any contractual obligation with third-party suppliers, and as with the procurement strategy mentioned earlier, policing should develop a decommissioning strategy that minimises operational impact while maintaining accountability robustness.

### *7.3.4 Learning as Underlying Theme to all Lifecycle Stages*

Each stage of the AI lifecycle will generate learnings, as indicated in the respective theme. These lessons should feed back into policing for when the next AI system enters the AI lifecycle. Optimally, lessons identified should be shared with other police and security forces to ensure a common and robust approach to AI.

## 7.4 AIPAS Accountability Themes and why Do the Themes Matter

Having established where and how accountability must be considered along the AI lifecycle, the chapter will now focus on how policing can best pragmatically operationalise accountability. Although the AI lifecycle provides a temporal structure, it does not itself specify the substantive accountability content. Procurement, design, deployment, adaptation, migration and decommissioning can only be managed accountably when they are tested against questions about the (1) AI system, (2) data, (3) laws and regulation, (4) risks, (5) oversight and redress, (6) stakeholders, (7) accountability evidence and (8) adaptability and learning.

These eight aspects or *themes* operate as the analytical framework that sits across the lifecycle. They provide the categories through which policing can identify accountability requirements, assign responsibility, evidence decisions and demonstrate that AI is being used lawfully, fairly, proportionately and safely.

### 7.4.1 The Purpose of the Accountability Themes

Introducing a new capability into policing, whether a computer system, vehicle, operational tool or AI system, involves multiple people, processes, approvals and responsibilities. A smooth transition relies on validated and proven methods of working. AI, however, presents a novel challenge in that it is a frontier capability, with limited precedent in policing, and forces are entering new territory in terms of procuring, developing, deploying, migrating and decommissioning AI systems. To support policing, we propose that accountability along the AI lifecycle should be viewed along eight paradigms that, when complete, offer the most robust framework for AI accountability in policing. The themes, which are displayed in Table 7.4, serve as foundational pillars for AI accountability and by grouping similar thematic areas together, makes what is arguably a complex environment to navigate, more intuitive and aligned to policing practices.

Informed by the principle-based AIPAS framework (Chap. 5), practical experiences and stakeholder validation (see Chap. 8), the themes are grounded in evidence-based research for practical application.

At first look, the themes in isolation may appear conceptually broad and abstract. However, for policing, they are specialist areas that often already exist, even if fragmented across different areas within a force. For instance, communication teams will be more familiar with the importance of stakeholders, whereas police IT specialists will be more familiar with aspects of data. By viewing the themes as a continuum and centralising them within AI accountability, policing can bring existing areas of expertise into a single framework structure. This allows accountability to be treated not in isolated disciplines but as a coordinated organisational responsibility that cuts across policing.

Policing should use the themes as a guardrail to understand where accountability is necessary, and where it needs to be demonstrated. Demonstrable accountability is important for policing; yet as in justice more broadly, it is not enough to simply do the right thing, but you must be seen to do the right thing [15] and be able to prove it.

**Table 7.4** AI accountability themes and their purpose in AI accountability

| *AI accountability theme* | *Purpose* |
|---|---|
| Theme 1: AI system | Defines the AI system's function, purpose, operational scope, limitations and the actors responsible for its development, deployment and oversight |
| Theme 2: Data | Focuses on the data used to design, train, test, deploy and operate the AI system, including its sources, quality, handling, governance and limitations |
| Theme 3: Laws and regulations | Ensures that the use of AI complies with applicable legislation, codes of practice, regulatory requirements and organisational policies |
| Theme 4: Risk assessment and management | Identifies, assesses and manages the risks, harms and biases associated with the AI system. This includes the risks created by using AI, the risks of not using AI and the mitigation measures required to protect fairness, operational effectiveness, public safety and public trust |
| Theme 5: Oversight and redress | Establishes mechanisms for oversight, independent scrutiny and redress, ensuring that AI systems, their outputs and related human decisions can be reviewed, challenged, corrected and learned from |
| Theme 6: Identification and management of stakeholders | Identifies and engages the stakeholders who are involved in, responsible for or affected by the AI system. This includes internal policing functions, external developers and suppliers, governance and oversight bodies, affected communities and members of the public |
| Theme 7: AI accountability evidence | Focuses on recording the decisions, processes, actions and evidence needed to demonstrate AI accountability |
| Theme 8: Adaptability and learning | Promotes continuous improvement through feedback, review, evaluation and organisational learning |

*Source* Akhgar et al. [1]; also Chap. 5

### 7.4.2 *Considerations per Theme*

Below, we provide recommendations how to address and implement each of the eight themes. The recommendations are based on the Implementation Guidance for AI in Policing and Security (the full guidance can be found in the Appendix).

#### 7.4.2.1 *Theme 1: AI System*

> The 'AI System' defines what the AI system does, how it functions, its purpose, limitations and who is responsible for its development and oversight. This helps ensure that the system is suitable for its stated aims, explainable to relevant audiences and aligned with legal and ethical standards. The better a police organisation details the AI system, the more effectively it can assess its appropriateness, identify potential challenges and implement safeguards ([1], p. 16).

The first theme addresses the AI system itself and is the most pragmatic starting point for policing. Before all else, policing must have a strategic vision of the system's function, intended purpose, limitations and individuals responsible for its procurement, deployment, migration and decommissioning. By articulating the AI system's purpose and actors involved (see Chap. 8), policing will have a better grasp of what accountability needs are required across the AI system and its lifecycle. The central principle is that an AI system should not be adopted solely due to availability; instead, adoption should occur when there is a defined operational need and when AI constitutes a justified and proportionate response to that need.

Like all themes, Theme 1 requires comprehensive documentation, particularly on:

- The decisions based around business need.
- Alternative options considered.
- Intended stakeholders.
- System scope.
- The contexts for both appropriate and inappropriate use.

Much of the foresight for this theme would come from the Chief Officer and enabling services. However, AI users must have input to ensure any AI system meets their requirements.

#### *7.4.2.2 Theme 2: Data*

The 'data' theme establishes an understanding of the data involved in creating the AI system (training/test data), as well as the data created for and by the AI system (inputs/outputs). It further ensures clarity about data management, risks and regulations, focusing on a number of critical considerations:

- Detailing the types and sources of data used in the development and deployment of the AI system, allowing proper evaluation.
- Establishing transparent procedures for lawful and appropriate acquisition, processing and management of data.
- Ensuring that legislative and policy compliance procedures in relation to data are effectively implemented and documented.
- Confirming the roles of responsible actors in relation to data are established, they are aware of their responsibility and possess appropriate skills and knowledge ([1], p. 20).

Data in policing is an already heavily regulated area, and for AI accountability, data requires particular attention due to the fact that data-driven decisions have real-world impacts and could hypothetically lead to wrongful lines of enquiry or, more seriously, wrongful arrests. The theme on data, therefore, requires policing to understand the types and sources of data used in developing the AI system, as well as the data used and generated during deployment.

It requires attention to be given to the whole data landscape, including:

- Data relating to marginalised or vulnerable groups.
- Any data bias.
- Personal data.
- How the data was collected and limitations of its use.
- Data processing and training.
- Data governance.
- Data storage.

Theme 2 also extends beyond types of data and into data appropriateness as data may be lawfully held, yet unsuitable for a specific AI purpose. For example, it may be suitable for analysis, but unsafe for automated or semi-automated decision-making. To understand the data, documentation is central to this theme and in some instances a legal requirement. Policing must also be proactive in their approach to data and not just do what is legally required, but also what is needed for accountability purposes, such as recording model cards, updates or any datasets which have been used. Data is also important during decommissioning/migrating. Not only for evidential reviews, but data that was generated during deployment may still be used in live case files (see Sect. 7.3.3 on decommissioning/migration).

#### 7.4.2.3 *Theme 3: Laws and Regulations*

Laws and regulations are an overarching principle in all areas of AI accountability. It ensures that the use of AI complies with applicable laws, codes of practice, regulatory requirements and policies. It further helps to clarify the legal context in which the system operates and safeguards public rights ([1], p. 25).

Theme 3 is unique as it extends across all the themes. However, it needs to exist as a standalone section because, arguably, it is the most important for policing. There are a myriad of laws and regulations that, context dependent, will influence police use of AI. These can include data protection laws, human rights laws, equality laws and more specific to policing, surveillance laws, criminal procedures, procurement rules and force policy.

Although policing is used to operating within heavily regulated areas, complexity arises when laws and regulations from different disciplines merge. This complexity can be demonstrated if an AI system wants to use data from highly regulated environments such as forensics, coupled with surveillance data and crime mapping intelligence. Each data source may have its own legal requirements, disclosure implications and oversight expectations. When brought together within a single AI system, the legal position may become more difficult to interpret. In the context of AI, laws and regulations must be continuously monitored, as an AI system may be lawful for one purpose with one category of data but not for another. Furthermore, the legal and regulatory landscape for AI is continuously changing, and new requirements may impact AI systems within policing. Therefore, Theme 3 requires that legal and regulatory compliance is an ongoing governance activity rather than a single event at the point of use.

#### *7.4.2.4 Theme 4: Risk Assessment and Management*

> This theme identifies risks, harms and biases associated with the AI system and implements mitigation measures. It covers both the impact of using and not using AI, with an emphasis on fairness and public safety. Identifying possible risk, harms and impacts of an AI system, as well as their likelihood is an important step to proactively manage these risks throughout the AI system lifecycle. It allows forward planning of how these risks (and associated harms) may be prevented, or where not possible, how they are to be managed and robust mitigation measures to be put in place ([1], p. 28).

Theme four addresses risk assessment and management and aligns closely with established police practices (ref-practical-accountability-chapter). While policing is familiar with risk management, AI systems introduce risks that may not be apparent through conventional operational assessments.

Risks when using AI can come from:

- Bias and heuristics.
- False positives and false negatives.
- Function creep (i.e. from misuse or over-reliance).
- Inadequate training or unclear responsibility.
- Erosion of public trust.

The theme also asks policing to consider the risk if they were not to use AI, which can be demonstrated through impact assessments and system reviews of what is currently best practice and associated friction points to the current practice. AI-based risk assessments require that risks be specified in detail, and generic statements such as 'bias may occur' or 'human monitoring will be used' are insufficient. An effective risk register must identify the risk, potential affected parties, likelihood, severity of harm, existing mitigation measures, mitigation ownership, residual risk and review timelines.

#### 7.4.2.5 *Theme 5: Oversight and Redress*

> 'Oversight and Redress' ensures mechanisms for independent scrutiny and AI accountability, allowing stakeholders, particularly the public, to challenge decisions and seek corrections, ensuring just outcomes. An appropriate oversight process ensures AI accountability through the effective monitoring and review of AI systems, their use, outcomes and impacts and thereby compliance and integrity throughout the AI lifecycle. An appropriate redress processes support AI accountability in allowing stakeholders, including the public, a clear means to challenge decisions, correct mistakes and seek justice. Redress helps build trust, ensure fairness and demonstrates that a policing organisation's actions are subject to review and correction ([1], p. 34).

Oversight and redress are an essential aspect to AI accountability because accountability becomes hollow without scrutiny and without redress that could amount to repeated failings. The theme identifies 'oversight' as the monitoring and review of AI systems, and 'redress' as the mechanism by which affected stakeholders can challenge decisions, correct errors and seek justice.

For Theme 5, and to answer possible oversight challenges, policing should be able to evidence at a minimum:

- Purpose and scope of the AI system.
- Data Model design and training.
- Deployment context.
- Decisions and outputs.
- User behaviour.
- Post-deployment monitoring.

- Governance processes.
- Risk register.
- Impact assessments.

Oversight and redress are a necessary theme because AI systems can obscure lines of responsibility, particularly because of the interlinked nature in how AI systems are created and deployed. An AI system may, for instance, be trained on inappropriate or biased data, applied to incomplete operational information and used by an officer who has not received sufficient training. If that system contributes to a wrongful arrest, accountability becomes difficult to locate, e.g. *does responsibility sit with the vendor, the organisation, the team, the supervisor, or the officer who acted on the output?*

Without clear oversight and redress arrangements, responsibility can become fragmented across the AI ecosystem, allowing each actor to point elsewhere when harm occurs. More importantly, affected individuals may be left without a clear route to challenge the decision, correct the error or seek justice. Oversight and redress therefore ensure that AI accountability is not only documented internally but is visible, reviewable and meaningful for those impacted by its use.

#### *7.4.2.6 Theme 6: Identification and Management of Stakeholders*

This theme aims at mapping and engaging all relevant parties who are involved in or affected by the AI system. This includes those internal to the policing organisation, external parties such as AI developers or oversight bodies and members of the public. A thorough identification and management of stakeholders ensures that:

- A full picture is available of which groups, bodies and organisations are responsible for AI accountabilities (e.g. oversight bodies, affected communities, AI providers) and what their respective roles are.
- A proactive management strategy for how to communicate and engage with these stakeholders in effective ways is in place

along with a clear strategy for how to respond to challenges, information requests or audit ([1], p. 40).

AI accountability requirements cannot be defined only by the organisation deploying the AI system, and as evidenced above, police use of AI can have an effect across entire communities. The identification and management of stakeholders require policing to identify all actors who deal with and are impacted by and who have an interest in the AI system.

Broadly, these are covered across the following groups:

- Police internal functions.
- Governance bodies.
- External organisations and expert groups.
- Specific communities and general public.

The need for the stakeholder theme is twofold: firstly, it allows policing to understand the accountability needs of other actors, and secondly, it allows policing to test whether their assumptions about risk, fairness, explanation and acceptability are correct. For example, a force may be satisfied that a system is legal, proportionate and technically reliable. However, that does not mean that communities will experience it as legitimate. Active stakeholder management enables clear communication, coordinated decision-making and early identification of concerns, and in turn, which will help build trust amongst the identified stakeholders.

#### *7.4.2.7 Theme 7: Accountability Evidence*

This theme ensures that evidence, about how AI accountability is assessed and captured, is robust, verifiable, accessible and understandable. This is not about criminal evidence but how policing evidences its AI accountability. By knowing how to prove AI accountability correctly, a police organisation will be better positioned to respond to challenges. The process will also highlight areas of AI accountability, which may need developing further ([1], p. 45).

The 'Accountability Evidence' theme is arguably the backbone of the entire accountability framework. In policing, accountability must be demonstrable, i.e. it is insufficient for policing to assert responsible conduct without evidence. Instead, policing must provide records, assessments, decisions, reviews and safeguards to substantiate that claim. The theme describes accountability evidence as documenting decisions, processes and actions taken to prove AI accountability. In turn, this will support requirements from internal reviews, external oversight and societal concerns. Policing already understands the importance of evidence in terms of investigations, and Theme 7 applies that same discipline to AI.

#### 7.4.2.8 *Theme 8: Adaptability and Learning*

Adaptability and learning ensures that policing stays up-to-date with technological innovations or requirements in AI and societal changes. It encourages continuous improvement through feedback, review and learning. It thus ensures that procedures and skills are reviewed and updated where necessary. Those developing or using the AI system remains aware of their responsibilities. Lessons are learned from mistakes, inefficiencies or unexpected/negative consequences ([1], p. 50).

AI systems used in policing cannot be treated as fixed tools, once they have been approved or deployed. The legal, operational, technical and social environment around AI will continue to change, and policing organisations need to be able to respond to those changes. This means that adaptability and learning should be built into the way that AI systems are governed from the outset. Therefore, to make best use of this theme, policing should not consider it as an afterthought but instead proactively manage and consider:

- Where along the AI lifecycle adaptability will be required.
- Development of a roadmap for lessons to be identified across the AI lifecycle to support learning (individual or organisational), e.g. on:
  - Procurement contracts.
  - Tradecraft.
  - The AI system.
  - Processes.
  - Management and governance.

For an AI system to remain safe, effective and accountable, it needs regular review across its lifecycle. This includes checking whether the system still meets its intended purpose, whether new risks have emerged, whether the data remains appropriate and whether changes in law, policy, public expectations or operational practice affect its continued use. Adaptability should also be reflected in relationships with suppliers, and where third parties provide or maintain AI systems. If possible, contracts should allow for review, modification, updates, suspension or withdrawal where necessary. This is particularly important where risks emerge after deployment, or where technical updates are required that were not anticipated in pre-deployment.

Although adaptability and learning is at the end of the theme structure, it is not the final step; it is the feedback loop that keeps the framework current. Adaptability and learning is therefore a continuous process, and as time passes and AI evolve, policing will be better positioned to manage an AI system with the next AI capabilities of the future.

## 7.5 Conclusion

As AI evolves and adapts, accountability must also be dynamic and iterative, continually addressing risks and societal concerns to maintain trust. At face value, the task may seem overwhelming particularly because AI and the use of AI across policing are a new and emerging enterprise. Strategic foresight must be applied when considering an AI system, and police leadership and enabling services must work closely together to 'bake in' accountability by design.

This chapter has demonstrated that pre-deployment is a critical accountability stage because decisions made during procurement and development shape how policing, and the AI system can demonstrate accountability. Deployment becomes the point at which accountability is tested in operational reality, requiring continuous monitoring, user feedback, public communication and the ability to change where risks emerge. Post-deployment demonstrates that the effects of an AI system may continue after it has been changed, transferred or decommissioned. Policing must therefore preserve any data needed to explain previous outputs and potentially redress human decisions which relied upon the AI systems output.

The eight accountability themes provide the substantive framework through which these lifecycle responsibilities can be made practical. The eight themes ensure that policing considers the whole accountability ecosystem from the AI system itself, the data it relies upon, the legal and regulatory environment, risk, oversight and redress, stakeholders, accountability evidence and adaptability and learning. Taken together, the lifecycle and themes provide a structured approach for turning high-level accountability principles into operational practice. Ultimately, AI accountability in policing must be expansive, demonstrable and iterative. It must be expansive because responsibility is distributed across a large number of stakeholders; it must be demonstrable because policing must be able to evidence how and why decisions were made and it must be iterative because AI systems, legal expectations, operational contexts and public attitudes will continue to change. This chapter offered practical considerations and recommendations on how to implement AI accountability in practice.

**Acknowledgements** Work in this chapter was partially funded under the AIPAS project ('AI Accountability for Policing and Security'; UKRI Engineering and Physical Sciences Research Council, Responsible AI (RAI) IA091 Grant Ref: EP/Y009800/1).

## References

1. B. Akhgar, P.S. Bayerl, F. Sampson, H. Gibson, M. Snowden, C. Rowley, A. Paradise, C. Spencer, J. Parkinson, *Guidance for Implementing AI Accountability in Policing* (AIPAS/CENTRIC Report, 2025)
2. D. Chalmers, *Post Office Horizon: An Accountability Failure* (Committee on Standards in Public Life, 2024, January 22). Retrieved from https://cspl.blog.gov.uk/2024/01/22/post-office-horizon-an-accountability-failure/
3. J.B.L. Chan, The technological game: How information technology is transforming police practice. Criminal Justice **1**(2), 139–159 (2001). https://doi.org/10.1177/1466802501001002001

4. J. Christie, The law commission and section 69 of the police and criminal evidence act 1984. Digit. Evidence Electron. Signature Law Rev. **20**, 62–95 (2023) Retrieved from https://journals.sas.ac.uk/deeslr/article/view/5642/5310
5. College of Policing, *Procurement Policy (Version 1.2)* (2025). Retrieved from https://assets.college.police.uk/s3fs-public/2025-04/Procurement-policy.pdf
6. D. De Silva, D. Alahakoon, An artificial intelligence life cycle: from conception to production. Patterns **3**, Article 100489 (2022). https://doi.org/10.1016/j.patter.2022.100489
7. European Parliament and Council of the European Union, Directive 2009/81/EC of the European Parliament and of the council of 13 July 2009 on the coordination of procedures for the award of certain works contracts, supply contracts and service contracts by contracting authorities or entities in the fields of defence and security, and amending directives 2004/17/EC and 2004/18/EC. Official J. Eur. Union, L **216**, 76–136 (2009) Retrieved from https://eur-lex.europa.eu/legal-content/EN/TXT/?uri=celex:32009L0081
8. European Parliament and Council of the European Union, Directive 2014/24/EU of the European Parliament and of the council of 26 February 2014 on public procurement and repealing directive 2004/18/EC. Official J. Eur. Union, L **94**, 65–242 (2014) Retrieved from https://eur-lex.europa.eu/legal-content/EN/TXT/?uri=CELEX:02014L0024-20180101
9. International Organization for Standardization, & International Electrotechnical Commission, *Information technology—Artificial intelligence—AI system life cycle processes (ISO/IEC Standard No. 5338:2023)* (2023). Retrieved from https://cdn.standards.iteh.ai/samples/81118/8e53a667079f4d959e93119fdbe84434/ISO-IEC-5338-2023.pdf
10. M.M. John, H. Holmström Olsson, J. Bosch, Architecting AI deployment: A systematic review of state-of-the-art and state-of-practice literature, in *Software Business: 11th International Conference on Software Business (ICSOB 2020), Karlskrona, Sweden, November 16–18, 2020. Lecture Notes in Business Information Processing*, ed. by E. Klotins, K. Wnuk, vol. 407, (Springer, Cham, Switzerland, 2020), pp. 14–29. https://doi.org/10.1007/978-3-030-67292-8_2
11. National Audit Office, *The Failure of the FiReControl Project* (The Stationery Office, London, England, 2011). Retrieved from https://www.nao.org.uk/reports/the-failure-of-the-firecontrol-project/
12. M.Y. Ng, S. Kapur, K.D. Blizinsky, T. Hernandez-Boussard, The AI life cycle: a holistic approach to creating ethical AI for health decisions. Nat. Med. **28**(11), 2247–2249 (2022). https://doi.org/10.1038/s41591-022-01993-y

13. Organisation for Economic Co-operation and Development, *OECD Framework for the Classification of AI Systems (OECD Digital Economy Papers No. 323)* (OECD Publishing, 2022). https://doi.org/10.1787/cb6d9eca-en
14. Procurement Act 2023, c. 54. Retrieved from https://www.legislation.gov.uk/ukpga/2023/54/contents
15. R v Sussex Justices, ex parte McCarthy, 1 KB 256 (1924)
16. D.P. Rosenbaum, A.M. Schuck, S.K. Costello, D.F. Hawkins, M.K. Ring, Attitudes toward the police: the effects of direct and vicarious experience. Police Q. **8**(3), 343–365 (2005). https://doi.org/10.1177/1098611104271085
17. South West Thames Regional Health Authority, *Report of the Inquiry into the London Ambulance Service* (1993). Retrieved from https://www.dcs.gla.ac.uk/~johnson/teaching/safety/reports/las.pdf

**Christopher Spencer** is Research Fellow at CENTRIC and a former Royal Marines Commando. His research interests sit at the intersection of security, intelligence, policing and artificial intelligence, with a particular focus on how emerging technologies can be used responsibly within operational security environments. He holds a BA (Hons) in International Relations and Intelligence from Staffordshire University.

**Petra Saskia Bayerl** is Professor of Digital Communication and Security and Head of Research at CENTRIC. Her research addresses societal implications and governance of new technologies in security and policing. She has been co-lead for both AP4AI and AIPAS and lead for national and EU-projects to improve the protection against online harms.

CHAPTER 8

# Practitioner Perceptions of What Is Needed for AI Accountability

*Christopher Spencer and Petra Saskia Bayerl*

## 8.1 Introduction

Accountability should sit at the centre of democratic governance and public decision-making. At its simplest, accountability requires individuals and institutions to explain, justify and where necessary, correcting their actions [13, 14]. Accountability is not a new concept and policing (see Chap. 2). As Markham and Punch [11] put it, 'policing *is* accountability', positioning accountability at the core of policing actions and responsibilities. However, the emergence of AI use within policing creates new accountability requirements that existing processes do not always fully address. In part, this is due to legislation and regulations that pre-date AI coupled with the long process it takes to update them, but also due to new challenges AI itself brings for policing and security applications (Chap. 1).

C. Spencer (✉) · P. S. Bayerl
CENTRIC, Sheffield Hallam University, Sheffield, UK
e-mail: christoper.spencer@shu.ac.uk



B. Akhgar and P. S. Bayerl (eds.), *Accountability for Artificial Intelligence in Policing and Security*, Security Informatics and Law Enforcement, https://doi.org/10.1007/978-3-032-41284-3_8

Current research on AI accountability has already indicated that responsibility can become difficult to locate when many actors contribute to the design, development, deployment and use of a system [8, 9, 17]. The AI accountability landscape now includes new actors like training data suppliers, test data suppliers, model developers and new decision factors that go beyond the human such as AI system settings, statistical weights and false positive and negatives. No longer will police need to only show how they (as humans) came to a decision, but also how the AI system came to a decision (or output), what influenced the decision and what influence the decision had. AI accountability therefore cannot be reduced to a question of whether an officer followed a policy or whether a force complied with a legal standard; instead, it becomes a wider question including how the AI ecosystem is governed, evidenced, explained, challenged and corrected.

This chapter addresses this operational question by examining expert perceptions on what is needed to make AI accountability meaningful within policing. Drawing on primary research conducted in the UK-based AI Accountability in Policing and Security (AIPAS) project with policing and security experts (Chap. 5)— and positioning that evidence alongside an international Accountability Principles for AI (AP4AI) expert consultation [1, 2, 4]—the chapter explores how AI accountability can—and should—be understood, demonstrated and operationalised in policing.

The chapter is informed by broader and ongoing work within the field of accountability for law enforcement [5, 7, 10, 12]. Although this body of research provides a well-established foundation for understanding police accountability, comparatively less attention has been given to the specific requirements of accountability for AI systems used in policing. This chapter addresses that gap by examining what policing consider necessary for AI accountability and the mechanisms needed to make abstract accountability operationally relevant for AI.

## 8.2 Methodology

Expert opinions were collected during 2024–2025 under the AIPAS project (Chap. 5). The research this chapter draws on consisted of a series of focus groups with 28 experts within the fields of forensics, counter terrorism (CT), serious and organised crime (SOC) and child sexual exploitation (CSE). The inclusion criteria for experts were foremost current operational knowledge and experience within one of these four policing fields. Further, as indicated by Hitt and Tyler [6], experiences, age, level of executive and risk orientation can influence decision-making. Therefore, the study recruited participants from a range of ranks/seniority to obtain greater diversity of perspectives. Along with rank/seniority, the experts also covered a variety of roles, including data specialists, policy advisers, data protection officers and policing practitioners. Initially, participants were recruited as an opportunity sample from the professional network of the researchers, then expanded via snowball sampling though participants' professional policing networks.

We conducted three expert consultations, each focused on a specific policing area (i.e. forensics, CT and combination of SOC and CSE). Each consultation session followed the same structure using a set of pre-defined questions (see Table 8.1). Data was both manually recorded in extensive notes and digitally through voice recordings. To ensure discussions remained grounded in the experts' field, one expert during each session provided a use case scenario. For forensics, AI accountability was framed around the use of AI in DNA and friction ridge detail; for CT, SOC and CSE, the use cases were centred around sized digital devices and 'big data' investigations. The purpose of the use cases was to frame expert discussions on where AI would sit in an investigation, the impact it would have in terms of accountability and what requirements and mechanisms would be needed to ensure accountability.

The discussions were structured along a prepared question set, including open and closed formats, that were captured either as part of discussions or as written answers (Table 8.1). The questions presented in each session were provided to experts in advance, along with details of the study and informed consent and information sheet.[1] The sessions were audio-recorded and transcribed verbatim to facilitate the analysis.

Analysis of the qualitative data started with open coding techniques allowing for the aggregation of key themes. Quantitative data generated through questions four and five (Table 8.1) was aggregated by adding together the responses across the sessions. This allowed the analysis to identify the overall ordering of the accountability principles for AI and

[1] The study obtained ethics approval from Sheffield Hallam University's Ethics Research Committee (Ethics ID: ER64540159). All data collected for the study abided by the regulations that are detailed within Sheffield Hallam University data storage and ethics and integrity policies.

**Table 8.1** Question set used during the expert consultations

| | *Question* | *Method of data capture* |
|---|---|---|
| 1 | What does accountability mean in your field? | Discussion |
| 2 | Do you know of any accountability and/ or AI-related frameworks in your field that we should consider for AI accountability in policing and security? | Discussion |
| 3 | Considering the 12 accountability principles for AI, are we missing any principles or are there any overlaps/ duplications, we should address? | Discussion |
| 4 | Please rank the 12 accountability principles for AI in order of importance for your field. | Ranking exercise |
| 5 | We have created a division of AI accountability principles into primary and supporting principles. Do you agree with this division at all and with the distribution of each principle into the two categories (i.e. primary/supporting)? | Yes/no questions with comment option |
| 6 | Considering the eight AIPAS themes and their sub-themes, we propose to assess AI accountability, are any themes or sub-themes missing or should any of them be merged? | Discussion |
| 7 | How do you demonstrate AI accountability to oversight bodies and public in this use case? Is this the same process you would use for other use cases in your field? | Open answers on answer sheet |

to compare how experts across the specialist areas prioritised them. The results were returned to experts for validation to ensure the analysis adequately represented their views. Once validated, the data was aggregated across the three sessions, and a comparative analysis was conducted to help identify similarities and differences across the specialist areas.

## 8.3 Results

### *8.3.1 Meaning of Accountability*

Accountability was described by the experts as a multi-layered concept involving answerability to the justice system, regulators and the public, as well as adherence to organisational standards, individual competences and the responsibilities of private providers delivering services to policing. Different specialisations within policing, however, offered different lenses in which accountability was viewed. Experts in the field of forensics linked accountability related to process and evidence assurance, i.e. accountability related to the courts and regulators employing system accreditation, professional competence, standard operating procedures, governance structures and audit trails. For CT experts, accountability was similarly framed through legal and regulatory mechanisms and professional ways of working. For experts in the fields of SOC and CSE, accountability focused more explicitly on victims, perpetrators and supervisors. The subsequent headings in this chapter group together themes where experts demonstrated their meaning on accountability.

#### *8.3.1.1 Organisational Answerability*

Across the expert groups' accountability was first understood as answerability. In forensics, this was expressed as organisational answerability to the justice system, courts, regulators and the public. In SOC and CSE, practitioners framed accountability through professional conduct, command structures and responsibility to victims, suspects, the public and courts. The responses reflect a broader policing culture in which decisions must be capable of being justified to multiple audiences, each with different expectations, rights and forms of scrutiny. AI complicates organisational answerability because responsibility may be shared across several actors. Policing may be accountable for adoption, procurement, deployment and oversight, but the supplier may control technical knowledge, while the practitioner may control the specific use and the supervisor may control authorisation and review. The international consultation [1] reinforces this distributed model by identifying an even wider range of groups that are relevant to AI accountability, including law enforcement agencies, judicial institutions, prosecutors, industry, independent technical bodies, civil society, universities, ministries, ombuds institutions and other public bodies. Together, these observations indicate that policing experts clearly viewed that accountability is a multi-stakeholder responsibility, with police one amongst many accountability actors.

#### 8.3.1.2 *Process and Evidential Robustness*

Experts repeatedly linked accountability to process and emphasised that accountability is tied to following the correct procedures rather than only to creating the correct outcome, meaning that the steps taken to reach a conclusion must be transparent and answerable across the entire chain of activity. This was particularly the case when experts were discussing evidential stands, and when considering what a chain of custody would involve if AI were used. The experts expressed that in order to fully understand what impact AI will have on judicial proceedings (i.e. AI outputs being used as evidence and how the evidence would be challenged), the experts would need to learn from a real court case.

Experts argued that evidential robustness for AI requires more than what is currently recorded in policing; when AI is involved, evidence would need to extend into records of the system used, such as AI configuration settings and data sources. This notion also aligns with experts from the AP4AI study, where accountability mechanisms identified included auditability of algorithms, data and design processes, internal and external auditing and user action logging.

#### 8.3.1.3 *Professional Conduct and Individual Responsibility*

The experts further reflected accountability through the lens of professional conduct and individual responsibility. In the ranking exercise, conduct was one of the higher-ranked principles across domains, alongside lawfulness, transparency, proof and explainability. In SOC and CSE, accountability was connected to professional conduct towards victims, perpetrators and the public, as well as to answerability to supervisors and chain-of-command for decisions made during and after an investigation. Forensics experts, in contrast, made a distinction between organisational answerability and individual responsibility by separating users who procure the tool from those who apply the tool, and by stating that practitioner-level accountability is centred on using the correct tool and correct methodology, including tool settings.

#### 8.3.1.4 *Legal, Ethical and Public Justification*

Experts placed lawfulness at the centre of AI accountability and across the fields connected accountability to compliance with laws, regulations and policy. This view reflects the basic policing requirement that power must be exercised lawfully, proportionately and in accordance with public expectations. However, discussions also showed that lawfulness alone was not considered sufficient to ensure accountability. Forensic experts emphasised public accountability and policing by consent and identified challenges in explaining AI-processed forensic evidence to citizens, juries and judges in a way that supports understanding and trust.

#### 8.3.1.5 *Auditability, Review and Challenge*

The final aspects identified were a combination of auditability, redress mechanisms, independent governance and oversight bodies as recurring accountability components. Forensics experts associated accountability with documented processes, evidence of reasoning behind decisions and the ability to explain outcomes to judges and juries. In the CT sessions, experts made the argument that if an outcome is successfully challenged by a defence, an accountability mechanism may already have failed earlier in the process. This is a significant insight because it treats courtroom challenge as a late-stage symptom of earlier process weakness, rather than as the primary mechanism of accountability. Effective AI accountability should therefore be upstream and anticipatory to possible legal challenges.

### 8.3.2 *What Experts Need for AI Accountability in Policing*

AI accountability can cannot exist in a vacuum; it relies on mechanisms, frameworks and a common understanding across individual officers or staff members through to the wider organisations and their stakeholders. Functional AI accountability therefore needs to be accessible to all, reduce ambiguity in responsibility and provide actionable guidance for decision-making purposes. The expert findings below consequently indicate that meaningful AI accountability depends on a combination of factors that, when taken together, offer robust AI accountability requirements that can meet the demand for policing contexts.

### 8.3.3 *AI Accountability Requirements*

#### 8.3.3.1 *Clear and Accessible Accountability Language*

Experts emphasised that terms need to be immediately understandable to a policing audience. In the discussions with experts the three principles of completeness, inclusivity and compellability emerged as particularly difficult terms to understand without prior knowledge of AI accountability. Experts in CT, SOC and CSE argued that transparency and explainability were closely linked and felt that in practice, it may be difficult to distinguish between them unless their definitions were made clearer. Experts from forensics also suggested that the two principles could potentially be condensed (transparency and explainability), although concerns were then raised that combining them may make the broader category harder to measure or operationalise.

#### 8.3.3.2 *Alignment with Existing Policing Systems*

Across the sessions, experts made reference to various policing systems to ensure accountability, both in specific and general terms, which are indicated in Table 8.2. Systems in this sense refer to frameworks, standards and decision models used in policing that offer accountability, rather than machine/AI systems. Experts indicated that it would be necessary for any AI accountability framework to be aligned with pre-existing systems, for two main reasons: firstly, it ensures that duplication of effort is reduced (i.e. not needing to complete a Data Protection Impact Assessment if there is already one in place); and secondly, it ensures that any framework is fit for purpose and does not contradict or work outside of other systems.

#### 8.3.3.3 *Evidence-Based Records Across the AI Lifecycle*

A repeated requirement voiced across the sessions was for adequate evidence, in the sense that accountability can and needs to be demonstrated. Accountability is not satisfied by stating that a system is lawful, fair or accurate; instead, it also calls for technical and organisational evidence that can be inspected. The experts across the sessions identified specific evidence needs in terms of transparency and explainability (see below) and particularly in relation to data, processes, outcomes and record keeping. All experts noted that accountability evidence would need to be a cross-cutting requirement across domains and across the full lifecycle of an AI system.

#### 8.3.3.4 *Transparency and Explainability Matched to Different Audiences*

Experts consistently identified transparency and explainability as central but also recognised that they cannot be understood as only type of disclosure requirement. Forensics participants stated that transparency

**Table 8.2** Current accountability systems used in UK policing

| *Data* | *Assessments* | *Frameworks and code of practice* | *Accreditation* | *Decisions and agreements* |
|---|---|---|---|---|
| General data protection rights | Risk assessments | Ethics frameworks | International Organisation for Standardisation (ISO) such as ISO 17020; ISO 17025 | Policy decisions |
| Data processing contracts | Human rights impact assessments | Police code of conduct | United Kingdom accreditation service (UKAS) | Senior investigating officer decisions |
| 'Big' data frameworks | Third-party assessments | Statutory code of practice | Other system accreditations | Security standards agreements |
| ALGOCARE | Data protection impact assessment | JPLANE model | Appropriate and current training | Procurement agreements |
| | | The science regulators code of practice | | National decision-making model |

and explainability are needed at all stages of the AI lifecycle, including design, development and deployment, but that communication must be balanced depending on the audience. CT experts similarly argued that the level of transparency and explainability must distinguish between what the public needs to understand and what police require, particularly where public disclosure could undermine techniques or sensitive capabilities. A moot point discussed, which could potentially create a gap in achieving transparency and explainability, was contractual agreements entered with third-party AI suppliers (as detailed in Sect. 8.3.3.6).

#### *8.3.3.5 Human Oversight and Defined Decision Ownership*

Human oversight appeared across the expert session as a requirement for trust, review and responsibility. Forensics experts discussed human-in-the-loop examples and indicated that human oversight is needed to support trust in the process, while also acknowledging that in some cases, systems may perform better than human recognition. CT, SOC and CSE experts described review mechanisms currently conducted by humans and distinguished between AI outputs used to direct an investigation and AI outputs used for evidential purposes. There was a clear distinction that—if any AI system was used in judicial proceedings—a human re-analysis would be required, not only to validate the output, but also to increase trust from a jury. Conversely, if an AI system's output supports information (e.g. AI used for initial translation), there would be lower requirements for human-in-the-loop intervention.

#### 8.3.3.6 *Supplier, Procurement and Third-Party Accountability*

Supplier and procurement accountability is an example of where accountability moves beyond policing (and emphasises the requirement for a broad stakeholder view discussed previously). Forensics experts recognised accreditation as a useful mechanism to ensure that accountability can be demonstrated; this approach would also assume responsibility for suppliers. However, in the case of a faulty AI system, strong contracts between policing and suppliers (that allow for redress) were suggested as an integral part of accountability. All experts, particularly within CT, identified difficulties where third-party providers may aim to protect commercially sensitive data, thereby creating obstacles to policing accountability.

#### 8.3.3.7 *Redress, Challenge and Corrective Mechanisms*

Redress is crucial because AI accountability without consequences is ineffective, and experts across fields expressed the need for mechanisms for redress, challenge and correction. It is worth highlighting that redress was positioned as a function mainly of independent governance and independent oversight structures, which may manifest as a complaints process or as auditing and reporting responsibilities. SOC and CSE experts expanded beyond supervisors and affected organisations and placed AI accountability within a stakeholder field that includes victims, perpetrators and the public. This national view aligns with international perspectives, which emphasised that for systems that produce harmful outputs or cannot be explained when challenged, there must be a route to stop, review, correct and remedy the issue [1].

#### *8.3.3.8 Training, Competence and Organisational Learning*

Training and learning emerged as central to practitioner perceptions of AI accountability. Forensics participants stated that learning and education are as important as training data, process and assurances. Similar to issues of evidence, experts positioned learning and education as a cross-domain accountability theme, indicating that practitioners cannot be accountable for AI use if they do not understand the tool, its limits, the relevant policy and the appropriate escalation routes.

### *8.3.4 Validating the Principles-Based Approach to AI Accountability*

Across the sessions, experts responded positively to the 12 AI accountability principles and broadly supported the principle-based approach as a meaningful foundation for strengthening AI accountability in policing and security. Responses indicated that the principles captured the issues practitioners regarded as essential.

Still, the 12 principles address very different aspects of policing accountability. We therefore aimed to understand whether certain principles were considered as more vital or 'central' to AI accountability. For this purpose, we asked participants to rank the principles according to their relevance to ensure AI accountability. The ranking results in Table 8.3 indicate that lawfulness was often ranked as the most important principle (average rank of 1.19). Transparency, conduct, proof and explainability were also ranked highly, suggesting that practitioners prioritise lawful use, openness, professional behaviour, evidential grounding and the ability to explain AI systems. At the same time, the ranking also demonstrates considerable differences across experts—indicated by the range between top (highest) rank and lowest rank per principle in Table 8.3. Explainability, for instance, was placed at 1 (most relevant), as well as 12 (least relevant), and a similar range can be observed for the majority of principles.

This observation is relevant, as it indicates that the relevance of principles is not fixed but highly variable, in that all principles can become 'most relevant' to demonstrate AI accountability depending on an expert's perspective or role, context, etc. It indicates that all 12 principles are required to ensure that AI accountability can be guaranteed in a cohesive way across AI deployment areas and instantiations.

**Table 8.3** Ranking results from across the three studies

| *Principle* | *Average rank* | *Top rank* | *Lowest rank* |
|---|---|---|---|
| Lawfulness | 1.19 | 1 | 7 |
| Transparency | 4.21 | 2 | 9 |
| Conduct | 4.32 | 1 | 8 |
| Proof | 5.52 | 1 | 10 |
| Explainability | 6.45 | 1 | 12 |
| Enforceability and redress | 7.12 | 2 | 10 |
| Learning | 7.12 | 2 | 10 |
| Compellability | 7.90 | 2 | 12 |
| Independence | 7.86 | 3 | 11 |
| Completeness | 7.91 | 1 | 12 |
| Constructiveness | 9 | 1 | 12 |
| Inclusivity | 9.25 | 3 | 12 |

#### *8.3.4.1 Measures to Achieve AI Accountability*

While the 12 AI accountability principles were seen as a solid conceptual foundation, experts argued that the principles alone would not guarantee AI accountability. The experts did not reject a principles-based approach; rather, they needed high-level principles to be translated into practical actions—an approach explicitly taken by AIPAS (see Chap. 5). AI was seen as such a new concept to policing that knowing what questions to ask may often be one of the greatest challenges to implementing AI accountability (i.e. there could be a scenario where accountability was missed not because of lack of due diligence, but because of inexperience). Experts were therefore rightly concerned with what must be recorded, explained, evidenced and reviewed and how. This question was put directly to experts in all sessions (see Table 8.2) with the request to record all currently existing or expected (wished-for) mechanisms to evidence the 12 AI accountability principles. Table 8.4 collates the ways and mechanisms provided by experts to demonstrate AI accountability against each of the 12 principles.

The findings from the experts offer several valuable insights. Firstly, there is a very clear need for AI accountability within policing. Other data-driven outputs such as intelligence and evidence are rightly scrutinised by stakeholders, and the same will need to happen for AI outputs to ensure that accountability standards (as a whole) in policing are maintained. Secondly, accountability mechanisms already exist across policing;

**Table 8.4** Expert feedback on mechanisms that can be used when attempting to demonstrate AI accountability

| *Principle* | *Accountability mechanism identified* |
|---|---|
| Lawfulness | • Clear legislation, new legislation or reviewed against current legislation<br>• Testing cases in court and consideration of legal challenges<br>• Compliance with: GDPR, case law, human rights act/ECHR, local force policy, policing codes of practice, data protraction act, Management of Police Information, freedom of information, equality act/equality duty<br>• Official accreditation or confirmation that a tool or use is not unlawful |
| Inclusivity | • Agreement for appropriate access for oversight or review<br>• Openness with the community about AI use<br>• Explanation of new developments<br>• Human validation where relevant and supported by risk plans, data management and retention plans<br>• Auditable and accurate records<br>• AI catalogue of records<br>• Standardised user guides and technical manuals |
| Completeness | • Communication plans for specific stakeholders<br>• Independent oversight supported by stakeholders<br>• Identification of relevant stakeholders and their needs<br>• Recognition of the stakeholder landscape (i.e. victims, perpetrators, the public, courts, supervisors, government bodies, public organisations, private organisations and anyone affected by police decisions) |
| Transparency | • Sharing information with internal stakeholders for validation, subject to security classification limits<br>• Privacy notices on projects<br>• Efficient freedom of information processes<br>• Annual reports from an AI body or commissioner<br>• Documented and auditable processes<br>• Correct management of information, including errors<br>• Accessible publication of system information<br>• Audience-specific information sheets for professionals, technical staff, users and the public<br>• Proactive communication about successful uses and failures<br>• Public consultations |

(continued)

**Table 8.4** (continued)

| *Principle* | *Accountability mechanism identified* |
|---|---|
| Impartiality | • Independent governance to monitor AI use<br>• An AI body or commissioner independent of police, security services and other operational bodies<br>• Independent auditing and oversight<br>• Monitoring and enforcement by independent organisations such as IOPC, an information commissioner-type body, HMICFRS or other scenario-dependent oversight bodies<br>• Governance boards<br>• Continual review and accreditation linked to independent monitoring |
| Proof | • Auditable processes and procedures where manual intervention is required<br>• Regular monitoring of manual intervention<br>• Evidence of use based on accreditation (such as ISO standards)<br>• Training and competence records<br>• Adherence to an 'AI code of conduct' similar to the forensic science regulator code of conduct<br>• Audit trails of actions taken<br>• Tamperproof and auditable records<br>• Records of AI testing<br>• Accreditation to industry standards<br>• Evidence that the system meets criminal evidential standards where relevant<br>• Pre-established requirements for providing records to court or review bodies such as HMICFRS<br>• Documentation and auditing supported by standard operating procedures, professional standards, policy and guidelines |
| Enforceability and redress | • Independent governance or oversight structures<br>• Oversight by an independent body or commissioner<br>• Independent escalation arrangements similar to an ICO-type body<br>• Forensic science regulator involvement where relevant<br>• Governance structures and reporting management information<br>• Clear police complaints processes<br>• Independent external agencies tasked with review and regulatory action<br>• Clear consequences for breaches of standards<br>• Legislation or contractual arrangements that include AI manufacturers within redress requirements where relevant |

(continued)

**Table 8.4** (continued)

| *Principle* | *Accountability mechanism identified* |
| --- | --- |
| Compellability | • Capturing associability requirements from all stakeholders<br>• Contracts, memorandum of understanding or equivalent agreements that specify details subject to regular review<br>• AI body or commissioner with powers over government AI users, including police and security services where relevant<br>• Full accreditation and monitoring by an independent body<br>• Mechanisms that allow oversight bodies to obtain the evidence needed to hold a force, supplier or system owner to account |
| Explainability | • Readily accessible reporting<br>• Engagement with AI developers<br>• Explanation of legal restraints by legal teams<br>• Equality assessment and policy communication<br>• Internal processes for evidencing dataset assurance<br>• Public consultation and education for policing, the public, the press and other audiences<br>• UKAS-style oversight of AI service delivery<br>• Court testing and translation of technical outputs into lay terms<br>• Training focused on the tool and its outcomes<br>• Simplified information or frameworks for different audiences<br>• Equality impact assessments<br>• Pop-it-style review (people, organisation, processes, information, technology)<br>• Case study examples<br>• Guidance and information sheets<br>• Clear explanation strategies based on the recipient's knowledge and constraints |

(continued)

**Table 8.4** (continued)

| *Principle* | *Accountability mechanism identified* |
|---|---|
| Constructiveness | • Working groups or boards engaging relevant parties<br>• Communication with MPs, ministers, human rights groups, victim groups and the legal community<br>• Independent oversight similar to a DNA strategy board<br>• Consultation with relevant boards and expert groups<br>• Formal approval processes<br>• Governance or decision boards<br>• Independent review<br>• Public accountability boards<br>• Practitioner and super-user feedback<br>• Mandatory information sharing<br>• Business process mapping<br>• Continued involvement of the AI manufacturer after purchase |
| Conduct | • Regularly assessed standards<br>• Clear force guidelines for AI use<br>• Ethical approval where relevant<br>• Compliance with forensic science regulator standards<br>• Alignment with authorised professional practice, the National Decision Model, the JPLAN model or other operating procedure guidance, code of ethics and policing principles.<br>• Public scrutiny panels<br>• Conduct assurance through internal and external audit or review |
| Learning | • Robust training with ongoing refresher assessment<br>• Feedback and new procedures from forces using AI<br>• Engagement with NPCC leads and strategic partners<br>• Records of improvement actions<br>• Internal quality management systems<br>• Capture and sharing of lessons learned<br>• Cross-force streamlining<br>• Documentation of errors and anomalies<br>• Standardised improvement actions and root-cause analysis<br>• Formal continuous review and improvement processes<br>• Learning updates shared across networks<br>• Practitioner feedback mechanisms<br>• Integration into oversight boards<br>• Exploration of automatic feedback from AI systems<br>• Learning processes that can themselves be challenged or contested |

however, their relationship to AI is fragmented, and sometimes, a step removed from their initial purpose. This means that while some mechanisms offer a current 'best fit solution', they miss the intricacies that AI accountability demands. Thirdly, the accountability landscape in policing is extremely complex and when adding complex AI systems into complex landscapes, the likelihood of negative outcomes (such as fragmented reasonability, duplication, inconsistent practices) increase and therefore gaps in accountability will begin to appear. Lastly, accountability mechanisms sit at the strategic, tactical and operational level. Often this means that strategic-level mechanisms establish the wider governance environment and organisational expectations, while tactical-level mechanisms interpret into policy and procedures, and operational-level mechanisms execute the policy. Effective AI accountability therefore depends on clear alignment across all three levels and across all contexts (for example, generative AI, predictive AI and agentic AI will all have different accountability mechanisms when being used, particularly in terms of policy decisions).

## 8.4 Discussion

The expert evidence summarised in this chapter demonstrates that AI accountability within policing is layered, multidirectional and complex (also Chap. 2). This complexity becomes more pronounced through AI systems where accountability must extend beyond the individual user and into a wider AI accountability ecosystem (cp. Chap. 7). Existing work on AI accountability shows that responsibility can become fragmented when many people and organisations contribute to system outcomes [17]. Wider AI policy resear also shows that accountability must be understood as part of public governance, not simply as a technical or organisational issue [15]. Our study further demonstrates that AI accountability is highly context specific. While the overarching logic of the principle-based approach was proven to be useful and meaningful across all four policing contexts, the findings demonstrate that putting them into practice requires careful calibration within each context. Despite this complexity, AI accountability remains essential to policing by consent in the UK. Police legitimacy depends on public trust, and public trust depends in part on whether police decisions can be explained, justified, challenged and corrected [16].

Experts in our study argued that there are gaps in the current accountability arrangements, in so far as they are fragmented across roles and responsibilities and often do not account for AI-specific requirements. The gap is partly being bridged through existing processes, including DPIAs, accreditation, force policy, governance boards and contracts. However, none of these was designed specifically for AI and its wider ecosystem. The findings also demonstrate a knowledge gap within policing in that experts recognised that completing a comprehensive AI accountability assessment would be challenging without support, guidance and accessible language. This is important to consider when designing AI accountability into policing because a framework that only technical specialists can understand will not be sufficient for policing or for their stakeholders. AI accountability must be usable by police leaders, procurement teams, enabling services, operational users, oversight bodies and affected communities.

The central question then is not only what accountability means, but what AI accountability requires in practice? Indeed, there was much overlap in how different experts viewed accountability, but there were also subtle differences in how they perceived accountability and the mechanisms that they are governed by. For instance, an expert in the field of SOC will have no need to answer to 'ISO/IEC 17025-*General requirements for the competence of testing and calibration laboratories*' but may need, in the UK, a Professionalising Investigation Programme Level 2 (PIP 2) qualification. For accountability, this is useful evidence, as it demonstrates that AI accountability must be a framework that address all user needs, agnostic to the type of AI system being used or the field in which it is being used. This requires AI accountability to be extensive in terms of extending beyond operational use and all-encompassing, in the sense that it should be applicable to all forms of policing from administrative functions to operationally deployable systems.

The chapter is based on expert evidence from the fields of forensics, CT, SOC and CSE which are high-value domains for understanding AI accountability as they involve sensitive data, serious harm, public trust, evidential implications and specialist operational pressures. However, other domains, such as neighbourhood policing, public order and call handling, may create additional AI accountability requirements. Also, AI capabilities change rapidly, and accountability mechanisms or requirements may thus shift as AI systems become more autonomous, more integrated or more embedded into routine police workflows.

While we posit that the 12 principles are applicable across policing and security domains and 'future-proof' in the sense that they will stay valid even for emerging AI scenarios, the specific forms, applications and implementation measures of AI accountability will vary and will need to be updated to remain relevant.

The wider public has a legitimate role in shaping accountability, particularly in a policing-by-consent model, as it underlies UK policing. Public views, community expectations and affected stakeholder perspectives were not captured in this study; however, they were captured in a large-scale national citizen consultation on the police use of artificial intelligence [3]. This consultation demonstrates that AI accountability is a public expectation (Chap. 10), emphasising again the crucial nature to implement and evidence accountability in a way meaningful for the public.

The insights in this chapter outline how practitioners define accountability and its stakeholders, existing and potential barriers to the practical implementation of AI accountability, as well as the measures they expect to be in place to ensure AI is embedded in policing in an accountable fashion. These are an important foundation to develop meaningful approaches for the implementation of AI accountability that go beyond 'thinking tools' into operational practice and concrete actions. Other chapters in this book describe such practical approaches from organisational practices (Chaps. 6, 7) to software-aided solutions (9). A fully developed *Implementation Guidance* is provided in the Appendix. Together these solutions show that practitioners' expectations for practical AI accountability are imminently feasible, as long as they stay up to date with ongoing technical, legal and societal developments and conscious of practitioners' adapting needs.

**Acknowledgement** Work in this chapter was funded under the AIPAS project ('AI Accountability for Policing and Security'; UKRI Engineering and Physical Sciences Research Council, Responsible AI (RAI) IA091 Grant Ref: EP/Y009800/1).

## References

1. B. Akhgar, P.S. Bayerl, K. Bailey, R. Dennis, S. Heyes, A. Lyle, A. Raven, F. Sampson, M. Gercke, *Accountability Principles for Artificial Intelligence (AP4AI) in the Internal Security Domain: Summary Report on Expert Consultations* (AP4AI Report 1). CENTRIC (2022).

Retrieved from https://ap4ai.eu/sites/default/files/2023-09/AP4AI_Report1_27Jan2022.pdf
2. B. Akhgar, P.S. Bayerl, K. Bailey, R. Dennis, H. Gibson, S. Heyes, A. Lyle, A. Raven, F. Sampson, *Accountability Principles for Artificial Intelligence (AP4AI) in the Internal Security Domain: AP4AI Framework Blueprint* (AP4AI Report 2). Europol Innovation Lab (2022). Retrieved from https://ap4ai.eu/sites/default/files/2023-09/AP4AI_Framework_Blueprint_22Feb2022.pdf
3. P.S. Bayerl, B. Akhgar, F. Sampson, M. Snowden, C. Rowley, A. Paradise, C. Spencer, *Citizen Consultation on the Police Use of Artificial Intelligence (AI) in the UK: Acceptance, Trust and Accountability: Highlevel Report (Public Version).* CENTRIC (2025). Retrieved from: https://stwebsitesprodeu.blob.core.windows.net/centric-website/CENTRIC_AI+Policing_National-Citizen-Consultation(public-version).Oct2025-WEB.pdf
4. P.S. Bayerl, M. Obst, B. Akhgar, *International Citizen Consultation on AI Accountability in Policing (AP4AI Report 3)* (Retrieved from, CENTRIC, 2023) https://ap4ai.eu/sites/default/files/2023-10/AP4AI_International-Citizen-Consultation_October2023%28online-version%29.pdf
5. A. Corbo Crehan, R. Delaforce, B. Shipton, A new focus for accountability: enabling police to act well. Curr. Issues Crim. Just. **37**(4), 554–571 (2025). https://doi.org/10.1080/10345329.2025.2531316
6. M.A. Hitt, B.B. Tyler, Strategic decision models: Integrating different perspectives. Strateg. Manag. J. **12**(5), 327–351 (1991) Retrieved from https://www.jstor.org/stable/2486562
7. T. Jones, The accountability of policing, in *Handbook of Policing*, ed. by T. Newburn, 2nd edn., (Willan Publishing, Cullompton, England, 2009), pp. 693–724. https://doi.org/10.4324/9780203118238
8. O. Kudina, I. van de Poel, A sociotechnical system perspective on AI. Mind. Mach. **34**, Article 21 (2024). https://doi.org/10.1007/s11023-024-09680-2
9. J. Kurre: The Accountability, Responsibility & Governance as a Unified Strategy for AI (Doctoral Dissertation, National American University). (ProQuest Dissertations and Theses, 2024)
10. G. Lennon, N.R. Fyfe, J. McNeill, F. Sampson, Principles for accountable policing. Retrieved from (2023) https://www.police-foundation.org.uk/wp-content/uploads/2010/10/accountable_policing_FINAL.pdf
11. G. Markham, M. Punch, Embracing accountability: the way forward—part one. Policing **1**(3), 300–308 (2007)
12. P. Murphy, P. Eckersley, L. Ferry, Accountability and transparency: Police forces in England and Wales. Public Policy Adm. **32**(3), 197–213 (2017). https://doi.org/10.1177/0952076716671033

13. Organisation for Economic Co-operation and Development, *Accountability and Democratic Governance* (OECD Publishing, Paris, France, 2014). Retrieved from https://www.oecd.org/en/publications/accountability-and-democratic-governance_9789264183636-en.html
14. Y. Papadopoulos, *Understanding Accountability in Democratic Governance* (Cambridge University Press, Cambridge, England, 2023) Retrieved from https://www.cambridge.org/core/elements/understanding-accountability-in-democratic-governance/21D041C3604D7FCD4ECCCD75A299B56D
15. R. Paul, E. Carmel, J. Cobbe (Eds.), *Handbook on Public Policy and Artificial Intelligence* (Edward Elgar, Cheltenham, England, (2024). Retrieved from https://www.elgaronline.com/edcollbook/book/9781803922171/9781803922171.xml
16. G. Richards, Police accountability. Institute for Government (2022). Retrieved from https://www.instituteforgovernment.org.uk/article/explainer/police-accountability
17. S.C. Slota, K.R. Fleischmann, S. Greenberg, N. Verma, B. Cummings, L. Li, C. Shenefel, Many hands make many fingers to point: challenges in creating accountable AI. AI & Soc. **38**, 1287–1299 (2023). https://doi.org/10.1007/s00146-021-01302-0

**Christopher Spencer** is Research Fellow at CENTRIC and a former Royal Marines Commando. His research interests sit at the intersection of security, intelligence, policing and Artificial Intelligence, with a particular focus on how emerging technologies can be used responsibly within operational security environments. He holds a BA (Hons) in International Relations and Intelligence from Staffordshire University.

**Prof Petra Saskia Bayerl** is Professor of Digital Communication and Security and Head of Research at CENTRIC. Her research addresses societal implications and governance of new technologies in security and policing. She has been co-lead for both the AP4AI and AIPAS projects and lead for national and EU-projects to improve the protection against online harms.

CHAPTER 9

# Software-Aided AI Accountability

*Jordan Thompson and Babak Akhgar*

## 9.1 Why a Software-Aided Solution for AI Accountability?

AI initiatives in security and policing operate in a multi-role, multi-stage and multi-version environment. *Multi-role* means that the operational owner, technical lead or supplier, data owners, legal adviser and Data Protection Officer (DPO), information-security lead, oversight body and communications or community-liaison function each hold relevant knowledge. *Multi-stage* refers to the pre-deployment, deployment and post-deployment activities described in Chap. 7. *Multi-version* recognises that the AI system, its configuration and data, the assessment template and the supporting evidence may all change over time.

These characteristics make accountability a matter of organisational coordination rather than document completion. Decisions about purpose, necessity and data may begin during design or procurement, be tested

J. Thompson (✉) · B. Akhgar
CENTRIC (Centre of Excellence in Terrorism, Resilience, Intelligence and Organised Crime Research), Sheffield, UK
e-mail: jordan.thompson@shu.ac.uk



B. Akhgar and P. S. Bayerl (eds.), *Accountability for Artificial Intelligence in Policing and Security*, Security Informatics and Law Enforcement, https://doi.org/10.1007/978-3-032-41284-3_9

before deployment and require reconsideration when performance, use or configuration changes. Evidence may sit in evaluation reports, supplier documentation, policies, minutes, risk records and operational logs. Unless an organisation connects those materials to named decisions and owners, it becomes difficult to establish what was known, who reviewed it and why a course of action was authorised. Software can provide that connective structure, while the substantive judgements remain with the responsible people and governance bodies.

### 9.1.1 Challenges of Traditional Approaches

Paper forms, shared folders, email chains and ad hoc spreadsheets can capture individual contributions, but they generally do not represent the full accountability environment. A static form does not reliably allocate questions to the function with the relevant authority; a legal-basis question, for example, may be completed by whoever happens to hold the document rather than reviewed by the appropriate legal or data-protection role. Concurrent copies also obscure which version is current, separate comments from the answer they concern and make the history of a decision difficult to reconstruct.

Traditional formats are also limited when the relevance of later questions depends on earlier answers. Without conditional routing, teams may collect material that does not apply or omit evidence that a particular deployment requires. Links between an answer, its rationale, the underlying artefact and a follow-up action can weaken as files move or staff change. The result is not necessarily an absence of activity, but a fragmented record that is harder to review, compare and render consistently for internal assurance or appropriate external communication.

### 9.1.2 Benefits of a Dedicated Tool

A dedicated accountability tool can address these weaknesses in five related ways:

1. It translates principles and themes into structured questions, displaying relevant follow-on prompts through conditional logic
2. It places comments, assignments and resolution notes beside the answer under discussion, so that each function contributes within a shared record

3. It links rationales and evidence to stable assessment items and preserves a record of edits
4. It versioned templates allow requirements to be updated while retaining the basis on which earlier work was completed
5. The same governed record can support distinct reports for project teams, oversight bodies and where appropriate, public communication

Software support may also influence organisational practice by making outstanding actions and review history visible across roles. This visibility can reinforce accountability as continuous work rather than a one-off exercise before deployment. It also exposes costs that may be hidden by apparently inexpensive spreadsheets and shared documents, including repeated reconciliation, inconsistent assessments, duplicated template updates and the loss of organisational memory when staff change. A dedicated tool does not remove these risks, but it provides a controlled process through which they can be managed.

The benefit is continuity, not automated assurance. During design and procurement, the record can connect intended purpose, supplier claims, legal analysis, anticipated risks and evaluation plans. During deployment and operation, it can connect approved operating limits to monitoring, incidents, actions and material changes. During migration or decommissioning, it can preserve decisions about retention, disposal and lessons learned. In each case, the organisation must still decide what evidence is sufficient, who may accept risk and whether the proposed or continuing use is lawful, necessary, and proportionate.

## 9.2 What Is the AIPAS Tool?

The AI Accountability in Policing and Security (AIPAS) tool is an access-controlled web application through which security and policing organisations can assess and evidence accountability for a specific AI capability and use context. Its project spaces allow authorised participants to complete assessments, discuss questions, link evidence, record actions and produce reports. Access is currently provided as a controlled access rather than as a public self-service application.

The application combines a core 'Self-assessment' with a collaborative interface, 'Check Assistant', 'Knowledge Base' and 'Template Editor',

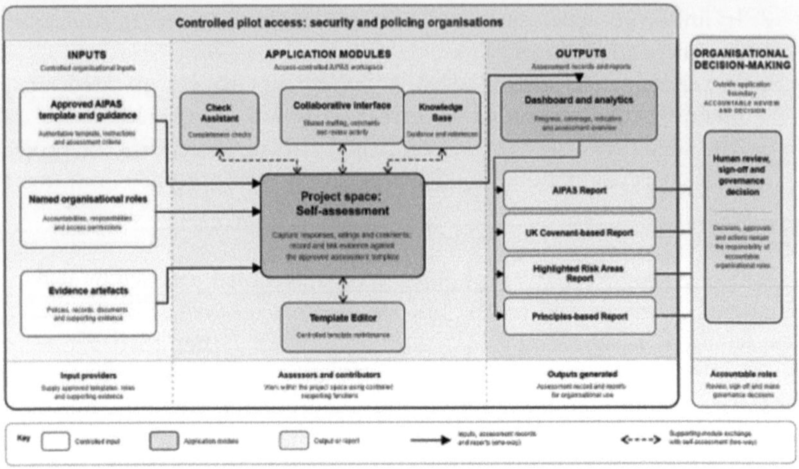


**Fig. 9.1** AIPAS module map, showing how controlled inputs and supporting functions contribute to an assessment record and its reports while review, sign-off and governance decisions remain organisational responsibilities

while project spaces, analytics and reporting organise the resulting material at programme level. These functions connect approved questions and organisational evidence within a reviewable assessment record. Figure 9.1 shows that relationship and places the resulting governance decision outside the application boundary.

### 9.2.1 *Self-assessment*

The 'Self-assessment' is the core function, and each assessment concerns a specific AI capability and use context rather than AI in the abstract. Its questions connect practical information to the twelve principles set out in Chap. 5: lawfulness, proof, completeness, inclusivity, explainability, transparency, compellability, enforceability and redress, impartiality, constructiveness, conduct and learning. Short hints explain why a question matters and indicate when supporting evidence is expected, while conditional routing limits follow-on questions to the circumstances recorded in the assessment.

Each applicable item uses one of three assessment states: 'Achieved' records the organisation's assessment that the requirement has been met and should be supported by an adequate rationale and evidence; 'Needs addressing' identifies a gap requiring an owner and follow-up and 'Not applicable' records that the item does not apply in the stated context and requires an explanation where the reason is not self-evident. These states structure review rather than constituting an automated finding of compliance. Useful prompts include what information will be communicated about an AI system's purpose and limitations, and what complaint and redress routes are available to affected people.

### 9.2.2 *Collaborative Interface*

Each assessment question has a linked discussion thread within which authorised participants can comment, reply, link an artefact, identify a colleague, assign follow-up work and mark a discussion as resolved or reopen it. The author and time attached to each contribution preserve the relationship between clarification and decision: for example, an operational owner may draft the intended purpose, a DPO may query retention, a supplier may link an evaluation report and an oversight member may request a clearer redress route. Keeping that exchange beside the answer reduces separate email chains and helps later reviewers understand how an issue was closed.

### 9.2.3 *Check Assistant*

The 'Check Assistant' is an embedded AI-supported aid for reviewing free-text quality. It examines three matters: whether an answer addresses every part of the question; whether the narrative conflicts with the question or selected assessment state and whether spelling or grammar may obscure meaning. Suggestions appear as non-destructive prompts that the author may accept or disregard.

The assistant does not decide whether an answer is correct, whether evidence is sufficient or whether an AI use is lawful or proportionate. Those judgements depend on facts and authority that the software may not possess, including non-public policy, operational context and legal advice. This boundary also limits automation bias: a drafting aid must not acquire the status of an accountable reviewer. Responsibility therefore remains with the named operational, legal, technical and oversight roles.

### 9.2.4 Knowledge Base

The 'Knowledge Base' supports interpretation of the assessment by answering natural-language questions against curated AIPAS guidance and returning grounded excerpts with links to the relevant page, so that the source can be read in context. Because it is bounded to that material and does not search the wider internet, its function is to make agreed definitions and guidance easier to locate rather than to replace professional advice or organisational policy.

### 9.2.5 Template Editor

The 'Template Editor' manages the structured set of stages, questions, input types, hints and warning tags used by the 'Self-assessment'. Templates are versioned using semantic versioning. A core template can act as the parent of a local child template, enabling an organisation to add policy text, questions or stricter thresholds without overwriting the core.

The template design allows a new version to be published without interrupting an assessment already under way. Where an upgrade is available, the guided migration process is designed to identify added, changed and removed items, retain archived answers for removed questions and record the time, user and summary of the change. Core changes can therefore be applied to related assessments while local child changes remain intact. This approach helps prevent multiple competing questionnaires and preserves which requirements applied when an assessment was undertaken; it does not remove the need for authorised template governance.

### 9.2.6 Project Spaces, Analytics and Reporting

Project spaces group assessments by capability or programme and restrict participation to authorised roles. Dashboards summarise the distribution of 'Achieved', 'Needs addressing' and 'Not applicable' responses by assessment and theme and can display change over time. These views direct attention to recorded gaps; they are not a substitute for examining the rationale and evidence beneath each state.

The AIPAS tool produces four report forms. The 'AIPAS Report' presents the full assessment for internal review. The 'UK Covenant-based Report' is specific to UK policing and organises relevant material against the National Police Chiefs' Council's Covenant for Using Artificial Intelligence in Policing. The 'Highlighted Risk Areas Report' draws together items marked 'Needs addressing' for action and review. The 'Principles-based Report' reorganises assessment material around the twelve accountability principles. Section 9.7 examines the content and intended audiences of these outputs in greater detail.

### *9.2.7 What the AIPAS Tool Is Not*

The AIPAS tool is not an automated ethics, legality or deployment decision-maker: it cannot verify all evidence, accept organisational risk, confer compliance or replace a Data Protection Impact Assessment, independent scrutiny or an authorised sign-off process. It provides structure for evidence and deliberation, but the organisations and people using that structure remain responsible for the decisions taken, and access to the controlled access does not itself validate an AI system or its use.

## 9.3 Roles and Responsibilities

The AIPAS tool is intended for multidisciplinary use, but software permissions do not create or transfer legal and organisational responsibility. The titles below describe an applied UK security and policing arrangement rather than a universal model; other security and policing organisations should map the same functions to their own lawful governance and oversight structures.

- **Project Lead or Operational Owner, Including a Senior Responsible Owner (SRO) Where That Title Applies**: establishes the operational need, records necessity and proportionality, secures resources, owns accepted risk and ensures that the authorised body records review and sign-off [5].
- **Technical Lead or Supplier**: documents architecture, data flows, training and testing, performance boundaries and security controls and provides verifiable artefacts such as evaluation reports, model or system cards and data-provenance summaries. Supply-chain information must remain current enough to support lawful processing and change control [3].

- **Legal Adviser and Data Protection Officer (DPO)**: advise on lawful basis, transparency duties and controller–processor relationships and lead or assure the Data Protection Impact Assessment. Their involvement should begin early and continue when the context changes [2].
- **Ethics or Oversight Body**: challenges necessity, proportionality, fairness and redress before deployment and at appropriate points in operation. Minutes and, where appropriate, public summaries should preserve the basis of that scrutiny [1, 4].
- **Information-Security Lead**: assures security architecture, resilience, logging and monitoring and confirms that material model, prompt or configuration changes follow auditable change governance integrated with data protection and oversight [3].
- **Public Communications or Community Liaison Function**: prepares plain-English transparency material, coordinates appropriate engagement and supports lawful responses to information requests with the operational owner and Data Protection Officer [1, 6].

Local AI and data policies, board terms of reference and delegations determine how these functions interact and who may approve a decision. The AIPAS tool should reflect those arrangements through access, assignments and review records rather than establishing a parallel governance structure.

## 9.4 Assessment Structure: Sections and Logic

Each AIPAS assessment has two top-level parts, with conditional questions within them: 'General information' and the 'AIPAS Accountability assessment' (see Fig. 9.2).

### *9.4.1 General Information*

'General information' establishes the intended purpose and use context of the AI capability, the organisation's role in its provision or deployment and the recorded assessment of whether the EU AI Act applies and whether the proposed use is prohibited or high-risk. These are organisational determinations, not classifications made by AIPAS (for legal context see Chap. 4). The section also asks whether third-party or pre-trained models are involved. Where they are, conditional questions

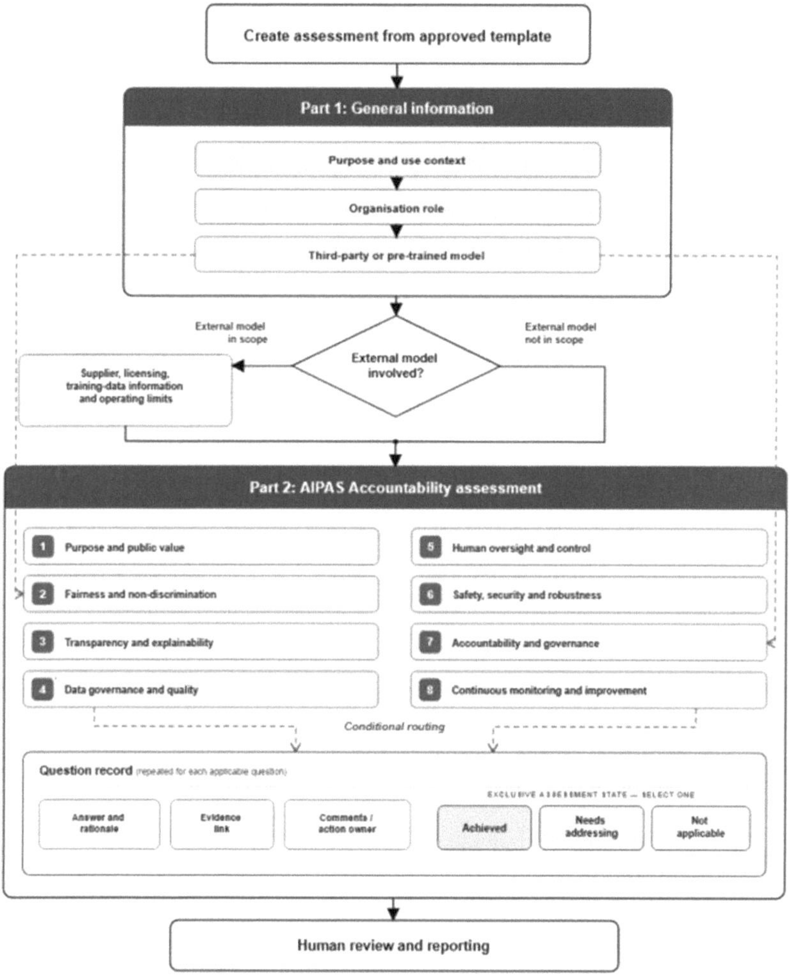


**Fig. 9.2** Two-part AIPAS assessment structure: general information controls conditional routing into the eight accountability themes, where answers are evidenced and recorded as 'Achieved', 'Needs addressing' or 'Not applicable' before human review and reporting

capture supplier details, licensing, available information about training data and stated operational limitations. This contextual record controls which later questions are displayed.

### 9.4.2 *AIPAS Accountability Assessment*

The second part collects answers, rationales and evidence across the eight practical themes (as defined in Chap. 5):

1. **AI System**: intended purpose, architecture, responsible actors, decision records and operating limits
2. **Data**: sources, types, acquisition and management, controllers and processors, quality and control
3. **Laws and Regulations**: applicable law, policing powers, policy and the means by which obligations are addressed
4. **Risk Assessment and Management**: potential harms and impacts, mitigations, acceptance criteria, testing and evaluation
5. **Oversight and Redress**: oversight lines, independent review, complaint handling and available remedies
6. **Identification and Management of Stakeholders**: affected groups, engagement and consideration of societal interests and concerns
7. **Accountability Evidence**: artefact types, robustness, custody and retention
8. **Adaptability and Learning**: monitoring, material change, review and lessons across operation and decommissioning

Questions may require a free-text, single-choice or multiple-choice response. Conditional logic reveals only relevant follow-on prompts. For each applicable assessment item, the recorded state is 'Achieved', 'Needs addressing' or 'Not applicable'; no other assessment state is used. Rationales, evidence links, comments and assigned actions provide the material through which reviewers test that state.

Monitoring under 'Adaptability and Learning' must be specific enough to govern the deployed use. It should cover performance and operating limits, patterns of use and outcomes, incidents and redress and material changes to the model, data or configuration. The assessment should record named owners, thresholds or triggers for review and the action required when a threshold is crossed. The AIPAS tool organises these commitments and their evidence, while the responsible governance body determines whether the monitoring and response are adequate.

## 9.5 Using the AIPAS Tool Across the AI Lifecycle

The AIPAS tool supports assessment and evidence-gathering at successive lifecycle stages (see Fig. 9.3). It does not replace decisions taken by the operational owner, technical lead or supplier, legal adviser and DPO, information-security lead or oversight body; instead, it maintains a structured record from which those roles can review the basis for a decision, the evidence available at the time and any action that remains open (for details on AI lifecycle and accountability themes see Chaps. 5 and 7). The account below explains how the software supports their application.

- **Development and Design**: the assessment records the system's purpose and necessity, relevant dataset and system facts, planned tests and evaluation, identified risks and mitigations, the oversight model and the lawful basis and powers relied upon. The resulting artefacts are an initial assessment, an evidence register and an action list
- **Procurement**: the project team adds supplier evidence, records the obligations associated with provider and deployer roles, identifies requirements that must flow down through the supply chain and checks licensing and rights for third-party or general-purpose AI components. The resulting procurement pack is accompanied by a residual-risk statement and assigned supplier actions.
- **Deployment and Operation**: the assessment applies the monitoring scope defined in Sect. 9.4.2 and records its owners, thresholds and escalation routes. It also covers log generation, incident and redress processes, change control and the status of any applicable Fundamental Rights Impact Assessment (FRIA). Outputs include an internal operational accountability report and, where the authorised organisation decides to publish one, a public summary.
- **Migration and Decommissioning**: the assessment documents the off-ramp plan, data retention and disposal, the preservation of evidence and logs and final learning. It therefore supports a closure record, an archived report and a record of lessons learned. These records preserve organisational memory and can inform a later

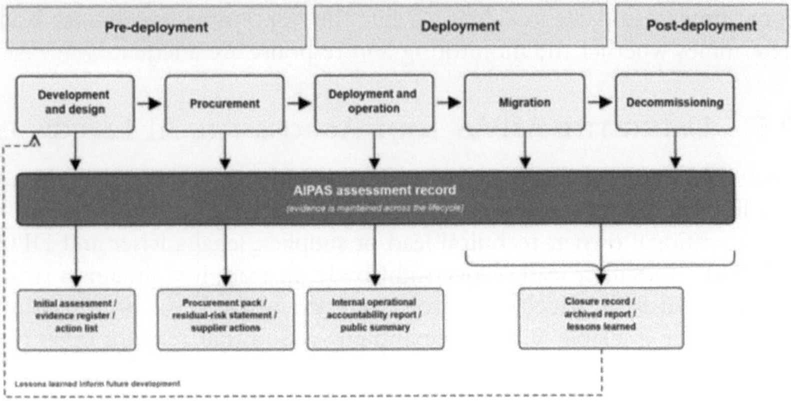


**Fig. 9.3** Use of the AIPAS assessment record across the AI lifecycle, from development and procurement to operation, migration and decommissioning

assessment; they do not make the software the owner of the underlying decision.

## 9.6 Collaboration, Projects and Access Control

Collaboration must preserve the relationship between each contribution, the assessment item it concerns and the organisational role responsible for the resulting decision. Contributors may act at different times, the assessment may be revisited as the system changes and both assessment and template versions must remain distinguishable. The AIPAS tool organises this activity within the operating environment described in Sect. 9.1 without transferring responsibility away from authorised people and bodies (see Fig. 9.4).

Organisations create project spaces to group related assessments, for example by capability, programme or deployment. Within an assessment, contributors can discuss a question in a threaded comment, record its resolution and retain the link to a stable question identifier. Role-Based Access Control separates the capabilities assigned to 'System Administrator', 'Organisation Administrator', 'Power User' and 'User', while multi-tenancy is designed to scope each document to an organisation. The application is also designed to record substantive actions, including

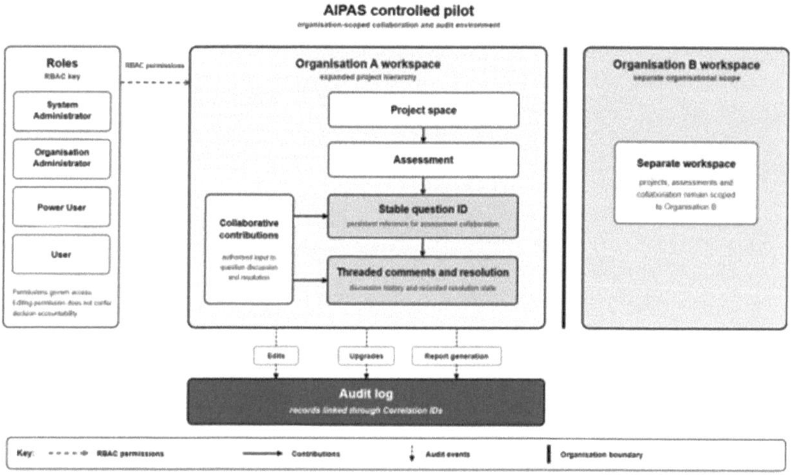


Fig. 9.4 Project collaboration, organisation-scoped workspaces, role-based permissions and audit logging in the AIPAS tool

edits, upgrades and report generation, in the audit log with a correlation identifier.

Administrators control participation in a workspace, while the relevant operational and governance arrangements determine who may supply evidence, review an answer or authorise a decision. Permission to contribute within the AIPAS tool does not, by itself, confer authority to approve an assessment or accept risk.

## 9.7 Evidence and Reporting

The AIPAS tool converts assessment records, whether complete or still being compiled, into audience-specific reports at any lifecycle stage. Each report is generated on demand from the current answers and template version and is watermarked with a timestamp, assessment identifier and template and version metadata, thereby preserving the basis from which it was produced (see Fig. 9.5). Reports can be viewed within the tool and exported as PDFs for governance packs and organisational records, while retaining the three assessment states defined in Sect. 9.2.1 rather than introducing a separate reporting scale.

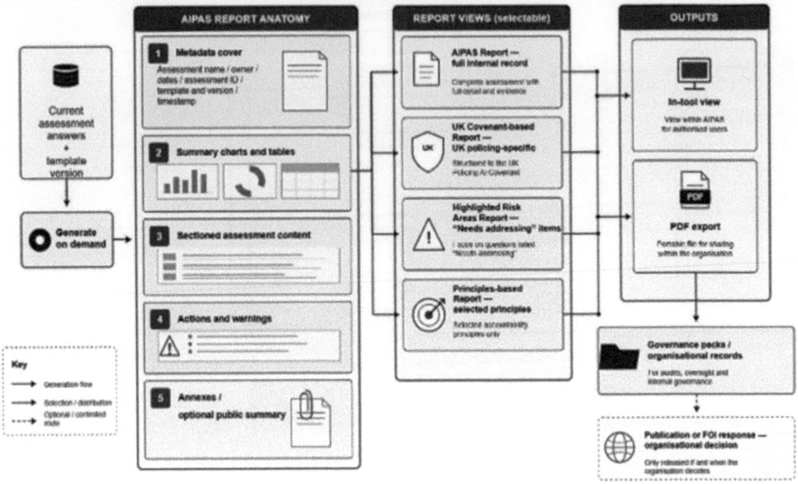


**Fig. 9.5** Anatomy and generation of AIPAS reports, showing traceable source metadata, four report views and controlled organisational outputs

Four formats serve different review purposes:

1. **AIPAS Report**: this is the full structured record for internal governance and audit. It covers purpose and scope, the data and system description, testing and evaluation, risks and mitigations, oversight and redress, security posture and the change log. Summary charts may show the distribution of items recorded in the three assessment states and changes in that distribution over time; they do not measure compliance or accountability.
2. **UK Covenant-based Report**: this UK policing-specific format filters relevant content and groups it under the National Police Chiefs' Council's *Covenant for Using Artificial Intelligence (AI) in Policing* (National Police Chiefs' Council, 2023). Inline callouts identify answers that relate to a Covenant requirement. It should not be presented as a general report for security and policing organisations outside the UK.
3. **Highlighted Risk Areas Report**: this view brings together answers marked 'Needs addressing' with their warnings and impacts. It can

identify material accountability risks, relevant UK Covenant requirements and missing documentation of oversight bodies or individuals, producing a focused record for action owners and governance review.

4. **Principles-Based Report**: this view filters the assessment by selected accountability principles, such as transparency, oversight and redress or learning. It enables procurement, legal and information-rights functions, scrutiny panels or communications teams to examine the material relevant to their remit.

Reports may be generated during design, procurement, pre-deployment review, operation and decommissioning. A PDF can contain cover metadata, sectioned content and annexes for governance or multi-agency use. The same structure may support a response to a Freedom of Information request, where appropriate, while enabling the responsible team to locate and redact sensitive annexes and preserve the public narrative. Summary charts and tables may show completion by theme, actions by owner and trends over time; an optional plain-English public summary may also be prepared. Decisions about disclosure, redaction and publication remain with the authorised organisation, not the software and the usefulness of these reports also depends on how the application and its assessment records are hosted, protected and maintained.

## 9.8 Deployment Options and Security Posture

The AIPAS tool is designed for deployment on an internal network or in a controlled government cloud, reached through approved organisational connections and configurable for restricted or air-gapped environments. Routine use is designed not to require external service calls, while the deployment architecture distinguishes the service that delivers the application from the service that holds assessment data. A deployment must use strong authentication and role-based permissions to restrict access to organisations, projects and functions, encrypt traffic and record substantive activity in structured audit logs. These controls make ownership and change control more visible but do not remove the organisation's responsibility for local security assurance, project membership, permission reviews, exported reports or action on audit findings.

The deployment model provides for automated build, test, security and quality checks before promotion to a test environment that mirrors

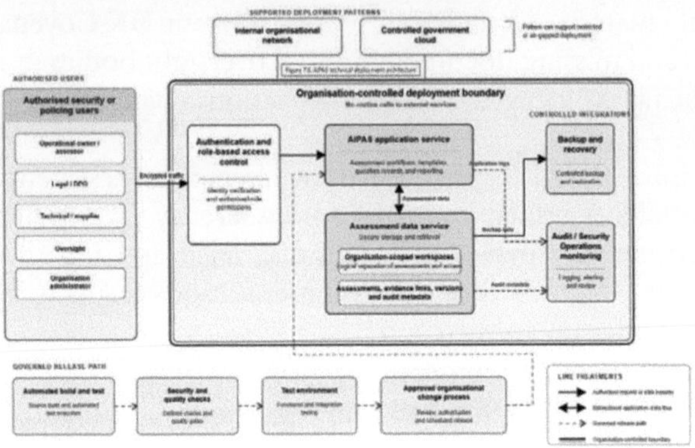


**Fig. 9.6** High-level AIPAS deployment architecture for controlled security and policing application, showing authorised access, organisation-controlled application and data services, workspace separation, audit and recovery integrations and the governed release path

production, after which releases should follow organisational maintenance and change-advisory arrangements that record what changed, when and by whom (see Fig. 9.6). The model is designed to scope assessment data to the relevant organisational workspace; administrators can configure encryption at rest, backup and recovery and connections to existing security monitoring, while autosave, concurrent-editing prompts and managed version changes support resilience. Each organisation must therefore assure its configuration against local security, data-protection, retention and recovery requirements, and this hosting description does not imply accreditation for a particular environment.

## 9.9 Operating the Assessment Process

Operating the AIPAS tool requires a defined routine for access, review, monitoring, version control and archiving because different functions contribute evidence at different points in the lifecycle, and each reviewed assessment version must remain distinguishable from later work. The application can support that routine, but accountable people must set the review points, evaluate the evidence and make the resulting decisions.

- **Accounts and Access**: organisation administrators should provision named users, assign the minimum permissions required for their role, manage project membership and review access when responsibilities change. Access remains limited to authorised participants in the controlled access. Project owners must also control any exported reports or evidence outside the application.
- **Assessment Timing and Versioning**: an initial assessment should begin when the proposed purpose and scope can be stated. It should then be reviewed during procurement, before deployment, during operation, after an incident that affects the assessment, following a material change and at migration or decommissioning. Before a major change, the team should preserve a snapshot of the existing assessment and create a new working version. Each answer should use one of the three assessment states defined in Sect. 9.2.1 and be supported by a rationale and, where the question requires it, durable evidence.
- **Review and Monitoring**: the project should establish scheduled governance reviews and identify the responsible participants in advance. Comments and assignments can record clarification, challenge and resolution, but a closed comment does not itself demonstrate that a risk has been accepted or controlled. The monitoring commitments defined in Sect. 9.4.2 should be incorporated into this review routine by recording the owner, measure or evidence source, threshold, review frequency and escalation route for each element. A finding that crosses an agreed threshold should trigger review of the affected answers, risks, public information and decision record.
- **Finalisation and Archiving**: at each formal review point, the team should resolve or assign items marked 'Needs addressing', record the decision and its conditions, generate the required reports and preserve the assessment version with its evidence register, change and migration logs, comments and approvals. At decommissioning, the archive should also record retention or disposal decisions and lessons relevant to later security or policing projects.

## 9.10 Pitfalls and Anti-patterns

The AIPAS tool supports accountability only when embedded in effective governance. The recurring failures are connected: unclear ownership permits the wrong roles to answer, weak evidence practices make supplier claims or boilerplate appear sufficient and absent review triggers allow an initially adequate assessment to become detached from the deployed system. Other pitfalls concern the assessment boundary, oversight and redress, security, transparency, interoperability and decommissioning, each of which requires a response proportionate to the particular capability and use context. Table 9.1 consolidates these anti-patterns and the corresponding practices as a review aid, not as a substitute for analysing the deployment. The corrective practices in Table 9.1 can be operationalised through the following repeatable workflow.

## 9.11 AIPAS Assessment Workflow

The following eight steps provide a repeatable route from defining the assessment boundary to preserving the reviewed record. They are not a one-way sequence: evidence identified during challenge or governance review may require an earlier answer, role allocation or system boundary to be reconsidered and the full workflow should be repeated whenever the system, its context, or an accountability requirement changes.

1. **Establish the Project and Roles**: the organisation administrator should create the project, restrict access and invite the operational owner, technical lead or supplier, legal adviser and DPO, information-security lead and oversight body
2. **Define the Assessment Boundary**: the owners should create one assessment for each model or component with a materially distinct purpose, data flow, risk, deployment or governance route, grouping related assessments in the project
3. **Complete the General Information**: the team should state the intended purpose and operational context, the organisation's role in provision or deployment, the recorded position on EU AI Act applicability and risk classification and whether a third-party or pre-trained model is involved. These answers allow conditional routing to present the relevant accountability questions

**Table 9.1** Common pitfalls in AIPAS assessments and recommended practice

| *Pitfall* | *Typical manifestations* | *Underlying causes* | *Recommended practice* |
|---|---|---|---|
| Rubber-stamping the assessment | 'Achieved' lacks rationale; reports are not considered | Schedule pressure; compliance-only thinking | Review the live assessment; require rationale and evidence for 'Achieved'; examine every 'Needs addressing' item and record the decision |
| Wrong roles answering | Operational staff answer legal or security items; suppliers answer for deployers | Unclear ownership; absent reviewers | Pre-assign the operational owner, technical lead or supplier, legal adviser and DPO, information-security lead and oversight body; use comments for challenge; require owner-confirmed resolution |
| Assessments not kept current | The design snapshot ignores deployment, incidents or changes | No schedule; unclear triggers | Review at design, procurement, pre-deployment, quarterly during operation, post-incident and decommissioning; preserve, update, review and report after material change |
| Lawfulness recorded without monitoring or redress | A lawful basis exists, but no monitoring threshold, owner or complaint route | Legality is mistaken for accountability | Before deployment, define measures, operating limits, thresholds, owners, escalation and redress; track findings in later reviews |
| One assessment covers too many components | Pipelines, models or materially different deployments are combined | Duplication avoidance obscures ownership | Map the system of systems; separate assessments where models, data, risks, deployments or governance differ; group them in one project |

(continued)

**Table 9.1** (continued)

| *Pitfall* | *Typical manifestations* | *Underlying causes* | *Recommended practice* |
|---|---|---|---|
| Third-party model opacity | Data, limitations, licence terms, security or update cadence are missing | Brochure reliance; confidentiality concerns | Require and link model cards, data summaries, known failures, usage/licence terms, security information and update policy |
| Boilerplate answers | Generic text does not match local use | Copy-and-paste practice | Use the 'Knowledge Base' for definitions but provide project-specific rationale and local evidence |
| Oversight is added late | Ethics or scrutiny first reviews after deployment | Delay or sensitivity concerns | Engage oversight early; record questions, outcomes and actions; prepare a public summary once communication and redress routes are agreed |
| Poor evidence hygiene | Links break; material remains inaccessible or on personal drives | No evidence register or repository | Keep a register with durable, versioned locations and access restrictions; check links before review |
| Performance limits are ignored | 'High accuracy' lacks context; unsuitable conditions are absent | Optimism bias; vendor marketing | Record bounds, failure modes and 'do not use' conditions; link positive and negative evaluations; require verifiable artefacts |
| Weak change control | Model, data or configuration changes escape review; logs and summaries become stale | Informal updates; external copies; unclear ownership | Preserve, edit a new version, review and report; use the 'Template Editor' for question-set changes; reissue summaries and retain local branches and migration logs |

(continued)

**Table 9.1** (continued)

| *Pitfall* | *Typical manifestations* | *Underlying causes* | *Recommended practice* |
|---|---|---|---|
| Security is treated as accountability | ISO, CE or other control evidence is treated as sufficient | Over-reliance on security frameworks | Treat security as necessary but insufficient; apply equal rigour to stakeholders, oversight/redress, transparency, evidence and learning |
| No transparency or Freedom of Information plan | Public content is improvised; responses are slow or inconsistent | Transparency is a postscript | Draft the public summary during assessment; use the 'Principles-based Report' as a structured source; apply targeted lawful redaction |
| System boundaries are unclear | Scope omits upstream data or downstream decisions | No architectural map; organisational silos | Map the system and decision points; separate assessments where risk, model, data or governance responsibility changes |
| Decommissioning is neglected | No off-ramp; unmanaged retention; incomplete records | Deployment focus | Plan early; record retention/disposal; preserve evidence and logs; produce a closure report on withdrawal |
| Human factors are ignored | Wellbeing, workload, review capacity or training are absent | Technical dominance | Record wellbeing, workload/review assumptions, training and support under stakeholder and conduct questions |
| Supplier claims are treated as evidence | Marketing text is copied; 'proprietary' prevents review | Power asymmetry; urgency; confidentiality | Require test reports, audits or datasheets; record who reviewed protected material and where it is stored |

(continued)

**Table 9.1** (continued)

| *Pitfall* | *Typical manifestations* | *Underlying causes* | *Recommended practice* |
|---|---|---|---|
| Weak alignment with the National Police Chiefs' Council (NPCC) Covenant | Covenant gaps remain open in UK policing assessments | Tick-list treatment | For UK policing, run the 'UK Covenant-based Report' at design and formal reviews; hold unresolved 'Needs addressing' items unless a time-bound mitigation, owner and decision are recorded |
| Interoperability is overlooked | Outputs cannot be shared; partner audit trails break | Local optimisation; late partner input | Document partner needs, sharing mechanisms, legal bases and logging interoperability; include partners in review threads and preserve each decision record |

4. **Answer and Evidence the Assessment**: assigned owners should provide project-specific answers, select 'Achieved', 'Needs addressing' or 'Not applicable', explain the rationale and link durable evidence
5. **Challenge and Allocate Action**: participants should use per-question comments to record questions and responses, assign an owner and target date to every outstanding action and record why a thread is resolved rather than treating closure as approval
6. **Apply the Support Functions**: the team may use the 'Check Assistant' to identify incomplete, inconsistent or unclear writing and the 'Knowledge Base' to locate relevant AIPAS guidance, while retaining human responsibility for the accuracy, legality, proportionality and sufficiency of every answer
7. **Conduct the Governance Review**: the responsible governance body should review the evidence and all items marked 'Needs addressing', record whether work may proceed or must be held and document any conditions, accepted risks, owners and completion dates

8. **Report, Preserve and Repeat**: the team should generate the required internal report and, where appropriate, a public summary; preserve the reviewed version and evidence and begin a new version after a scheduled review, incident or material change

**Worked Example**: a policing team proposes an AI system to triage digital material. It creates a project and a separate assessment for the model, recording its investigative purpose, users, data and decision boundary. The supplier links an evaluation report and operating limits. The DPO marks retention 'Needs addressing' because the cited schedule does not cover the proposed data, while oversight asks for the panel's remit. The 'Check Assistant' identifies that the answer names the panel without explaining its authority; the answer owner adds the terms of reference and the reviewers record resolution. The governance body holds deployment approval pending an agreed retention position. Once evidenced, the team records the decision, generates the 'AIPAS Report' and preserves the reviewed version. A material model or data change would start a new version of the workflow.

## 9.12 Conclusion

This chapter has explained how structured software can support AI accountability in security and policing. The AIPAS tool brings questions, roles, evidence, discussion, version history and reporting into one assessment environment. Used across the AI lifecycle (Chap. 5), it can connect an operational decision with the evidence available and the people responsible for review, while preserving unresolved issues and later changes.

The value of that structure depends on disciplined use. An assessment state is not a decision, a completed comment is not an approval and a technically secure deployment is not the same as accountable use. Operational owners, technical specialists and suppliers, legal and data-protection advisers, information-security functions and oversight bodies must continue to exercise their own judgement and fulfil their respective duties. The AIPAS tool can organise and test the clarity of the record; it cannot determine lawfulness, proportionality, ethical acceptability or whether an AI system should be deployed.

The AIPAS tool is available only through controlled access for security and policing organisations, rather than as a public self-service application. The claims in this chapter are likewise limited to security and policing: no validation or transferability beyond those domains is asserted. Within those boundaries, the AIPAS tool can support a more consistent, reviewable and evidence-based accountability process, while leaving human, organisational and legal accountability with the institutions and individuals that hold it.

**Acknowledgements** Work in this chapter was partially funded under the AIPAS project ('AI Accountability for Policing and Security'; UKRI Engineering and Physical Sciences Research Council, Responsible AI (RAI) IA091 Grant Ref: EP/Y009800/1).

## References

1. College of Policing, Data ethics Authorised Professional Practice (APP): Consultation (2024). https://www.college.police.uk/app/data-ethics
2. Information Commissioner's Office, What are the accountability and governance implications of AI? (2023). https://ico.org.uk/for-organisations/uk-gdpr-guidance-and-resources/artificial-intelligence/guidance-on-ai-and-data-protection/what-are-the-accountability-and-governance-implications-of-ai/
3. Information Commissioner's Office, Governance and accountability in AI (Audit toolkit) (2025). https://ico.org.uk/for-organisations/audits/data-protection-audit-framework/toolkits/artificial-intelligence/governance-and-accountability-in-ai/
4. National Police Chiefs' Council, Covenant for Using Artificial Intelligence (AI) in Policing (2023). https://www.npcc.police.uk
5. Police Digital Service, & National Police Chiefs' Council, National Policing Digital Strategy 2025–2030 (2025). https://news.npcc.police.uk/resources/national-policing-digital-strategy-2025-2030
6. Warwickshire Police, Artificial Intelligence (AI) Policy (2025). https://www.warwickshire.police.uk/SysSiteAssets/foi-media/warwickshire/policies/a/artificial-intelligence-ai-policy.pdf

**Jordan Thompson** is a Research Fellow at CENTRIC and a specialist in the development of technical solutions for detecting physical threats, managing crises and responding to disasters. With a background in computer science, he has an academic focus on the intersection of technology and security.

**Babak Akhgar** is Professor of Informatics, Director of CENTRIC and a Fellow of the British Computer Society. He has extensive and hands-on experience in the development, management and execution of KM projects and large inter national security initiatives (e.g. combatting terrorism and organised crime, AI in Policing, cyber security, public order and cross-cultural ideology polarisation). He has more than 100 refereed publications in international journals and conferences on information systems, specifically focusing on knowledge management (KM). Babak was awarded an OBE in June 2022, the second highest order of the British Empire award, in recognition of his services to security research.

CHAPTER 10

# Societal Expectations for AI Deployments in Policing

*Petra Saskia Bayerl*

## 10.1 Introduction

The introduction of artificial intelligence (AI) for policing purposes continues to be a flashpoint in public discussions about both fairness and security. On the one hand, AI is seen as a necessary avenue to improve policing performance (e.g. in the context of community policing, counter-terrorism or cybercrime; [5–7]), and as such has become an integral part of future policing strategies (Chap. 1). On the other hand, legitimate concerns are raised about the potential of AI to make wrong decisions or

This chapter is an adaptation of Bayerl et al. [1].

P. S. Bayerl (✉)
CENTRIC (Centre of Excellence in Terrorism, Resilience, Intelligence and Organised Crime Research), Sheffield, UK
e-mail: p.s.bayerl@shu.ac.uk



B. Akhgar and P. S. Bayerl (eds.), *Accountability for Artificial Intelligence in Policing and Security*, Security Informatics and Law Enforcement, https://doi.org/10.1007/978-3-032-41284-3_10

ingrain historical biases. Extreme opinions even consider AI to be 'unfit for purpose' [3].

Still, AI deployments by police forces are ongoing, and it is therefore not a question of 'whether' but rather of 'how' to deploy AI in a way that is acceptable to the public. Yet, public perceptions of AI are not monolithic. Past research shows that public perceptions are split along varying viewpoints from acceptance to resistance (e.g. [2]), ranging from a perceived requirement for the better protection of society to the willingness to destroy surveillance infrastructure to safeguard one's privacy. Acceptance is further affected by the way that AI is used, such as levels of transparency, the non-sharing of data (i.e. privacy) and the presence of an appeals system [4]. The latter especially speaks to the importance of accountability for public acceptance of AI. Understanding variations in how the public perceives AI deployments in policing is therefore an important aspect to guide AI deployments and to shape AI policies in an effective manner.

This chapter offers insights from a large-scale national citizen consultation in the United Kingdom. It provides a detailed view not only about overall levels of acceptance but also where citizens see the most benefits and concerns in the AI use by police, as well as conditions that would improve acceptance and trust. The subsequent section outlines the survey methodology, followed by a detailed description of the findings. The chapter ends with recommendations for measures to increase public acceptance and trust, based on the evidence of the consultation.

## 10.2 Methodology of the Consultation

### *10.2.1 Data Collection and Participants*

The consultation was conducted as nation-wide anonymous online survey in the UK between April and June 2025, focusing on the general UK population starting from 14 years of age. The recruitment was conducted through an online panel provider and ensured representative sample demographics for gender and age (for age groups 18+) across all four nations in the United Kingdom.

In total, the consultation includes a total of 10,114 participants. Table 10.1 provides an overview of the sample characteristics. As the overview indicates, the consultation reached a diverse group of participants both in terms of demographics and experiences, which allows the investigation of subgroups within the sample.

### *10.2.2 Aspects Assessed in the Consultation*

The consultation covered seven themes to investigate different aspects of public perceptions and acceptance:

1. Acceptance of, and trust in, AI and AI-supported policing
2. Risks and benefits of AI for personal, societal and institutional policing purposes
3. Measures to improve public trust in the police use of AI
4. The role of accountability for trust
5. Public reactions to 32 deployment scenarios, including current and potential (future) deployment areas
6. Measures to recover trust in case of AI failures (open answers)
7. Trust in the police and other institutions

This chapter focuses on findings with respect to acceptance, risks, benefits and the role of AI accountability.

### *10.2.3 Ethics*

The study received ethics approval by the Sheffield Hallam University's Research Ethics Committee. The survey was conducted as an anonymous questionnaire (i.e. did not collect participant information such as location or IP address). Participants were further required to give informed consent before starting the survey (individuals who did not consent were not allowed to proceed to the survey); double consent was used for the youth sample, requiring parent consent before youths were able to participate in the consultation.

**Table 10.1** Sample characteristics

| *Category* | *Percentage (%)* |
|---|---|
| *Gender* | |
| Women | 51.9 |
| Men | 47.5 |
| Non-binary/other | 0.5 |
| Prefer not to say | 0.2 |
| *Age* | |
| 14–17 years | 5.7 |
| 18–24 years | 19.0 |
| 25–34 years | 16.6 |
| 35–44 years | 15.4 |
| 45–54 years | 16.0 |
| 55–64 years | 14.4 |
| 65–74 years | 9.4 |
| 75+ years | 3.5 |
| Prefer not to say | 0.1 |
| *Ethnicity* | |
| White | 80.1 |
| Black | 9.8 |
| Asian | 5.2 |
| Mixed ethnicity | 3.1 |
| Other | 1.5 |
| Prefer not to say | 0.3 |
| *Education level* | |
| No formal education | 0.6 |
| Secondary education | 31.7 |
| Bachelor/master degree | 34.4 |
| Professional degree | 7.9 |
| PhD | 4.7 |
| *Member of a group discriminated against by police* | |
| No | 78.9 |
| Yes | 11.2 |
| Unsure | 6.6 |
| Prefer not to say | 3.3 |
| *AI knowledge (general)* | |
| No or very little knowledge | 32.0 |
| Some or good knowledge | 65.0 |
| Expert | 3.0 |
| *AI knowledge (policing)* | |
| No or very little knowledge | 73.2 |
| Some or good knowledge | 25.4 |

(continued)

**Table 10.1** (continued)

| *Category* | *Percentage (%)* |
| --- | --- |
| Expert | 1.3 |
| *Professional background* | |
| Working in security-related sector (adult sample) | 6.3 |
| Parent working in security-related sector (youth sample) | 16.2 |

## 10.3 Findings

### *10.3.1 Extent of Public Support for AI Use by Police*

The consultation demonstrates a largely positive view about AI use by police: 61.3% of participants indicated support for the police use of AI; 18.9% of this group even voiced strong support. A similar percentage (64.0%) moreover considered AI as a capability that enhances policing (cp. Fig. 10.1). The reasons for support can be gathered from the free text comments provided at the end of the survey. These illustrate that support is founded in the expectation that AI makes current policing efforts more effective, as well as the expectation that AI will help the police react more effectively against future crime challenges. AI is thus seen as an important measure to 'futureproof' policing capabilities.

The police use of AI was opposed by 19.1% (7.2% of these indicating strong opposition), while 12.4% considered AI as a capability that undermines policing. Reviewing the free comments of these participants, the opposition was grounded in in a general lack of trust in the police, as well as fears that 'the police start seeing [individuals] as data rather than people'. Moreover, AI was seen as 'too premature' to be used for policing and security purposes.

Notable is the considerable number of people with a neutral stance (i.e. neither supportive nor opposing position). The neutral stance was similar in size to the group opposing AI use by police: 19.6% indicated a neutral position compared to 19.1% opposing AI.

An even larger group (23.6%) doubted whether AI has any effect (either positive or negative) for policing. Together, these observations indicate that a considerable majority of the public (80.9%) is not directly opposed to the police use of AI, although some may doubt whether it has any direct effects.

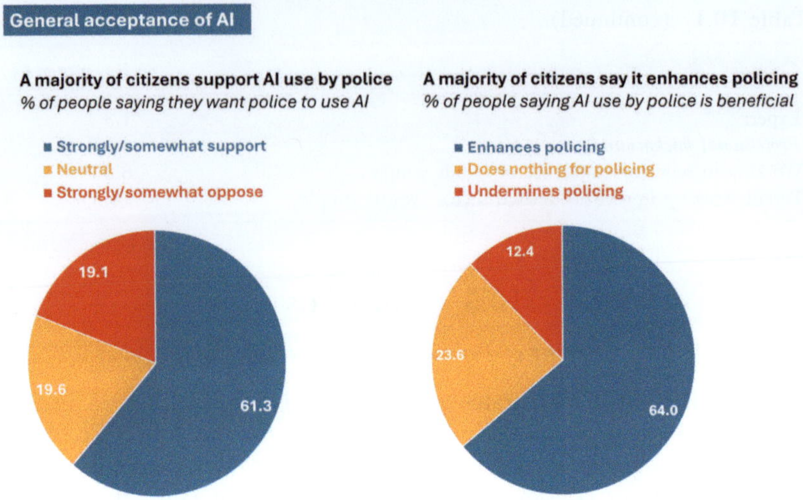


**Fig. 10.1** General acceptance of police use of AI (adapted from: Bayerl et al. [1])

**Subgroup Differences**: investigating subgroups within the sample identified relevant variations. For example, survey participants with higher levels of AI expertise indicated higher levels of acceptance: 88.1% of experts agreed that AI enhances policing, compared to 49.9% of people claiming no expertise in AI. Reversely, 33.9% of people without expertise believed that AI does nothing for policing, compared to only 6.7% of experts. Equally, participants with higher levels of education (PhD and professional degree) showed stronger acceptance: 77.8% and 77.1%, respectively, felt that AI enhances policing compared to 56.9% with secondary school as their highest attainment. While the exact links deserve further investigation, these observations indicate that knowledge and education attainment affect reactions to AI, potentially through greater confidence about the working and effects of AI.

Less clear are results for age groups and gender. Comparing answers in the youth and adult sample suggests that youths are somewhat more critical towards AI, with 59.8% agreeing that AI enhances policing and 26.2% seeing no effect, compared to the adult sample with 64.7% and 23.2%, respectively. Men were slightly more positive compared to women (men: 68.9% enhances policing, 20.0% does nothing versus women: 59.8% and 26.8%, respectively). A clearer effect was found for people identifying as non-binary with only 30.4% considering AI as enhancing policing (vs 32.6% undermining it). However, this group is extremely small (n = 46), which means that these numbers should only be seen as a potential indication for further investigation.

### *10.3.2 Main Concerns and Benefits*

To understand public views on concerns and benefits, we differentiated between three areas of impact:

1. **Personal aspects** addressing individual safety, privacy and the safety of the people someone cares about
2. **Societal aspects** addressing effects on communities or public goods such as fundamental rights and effective use of taxes
3. **Institutional aspects** addressing impacts on the ability of the police to fulfil their role

#### *10.3.2.1 Concerns*

The strongest public concern was the fear that AI could be used to escape the responsibility for failures, i.e. that 'police pass the blame to AI if something goes wrong' (cp. Fig. 10.2). This is followed by worries about errors and biases that will negatively affect specific groups. Less concerns were voiced about personal risks, such as individual privacy offline and errors and biases affecting oneself. The most pertinent public concerns therefore seem to be societal and institutional, rather than personal. A relevant observation is that concerns are affected by personal experiences. Perceived vulnerability to police discrimination affected concerns about the police passing blame to AI: 61.1% in this group were concerned or extremely concerned, compared to 48.7% of people who do not see themselves as vulnerable to police discrimination. Crime victimisation in the last 12 months also led to higher rates of concerns: 60.3% in this group voiced concerns about errors and biases that negatively affect specific groups compared to only 36.7% without crime experiences.

#### *10.3.2.2 Benefits*

The strongest benefits of AI were perceived for enhancing the effectiveness of the police to solve crimes and in an improved ability and more time for the police to deal with serious crimes. 67.8% and 66.4%, respectively,

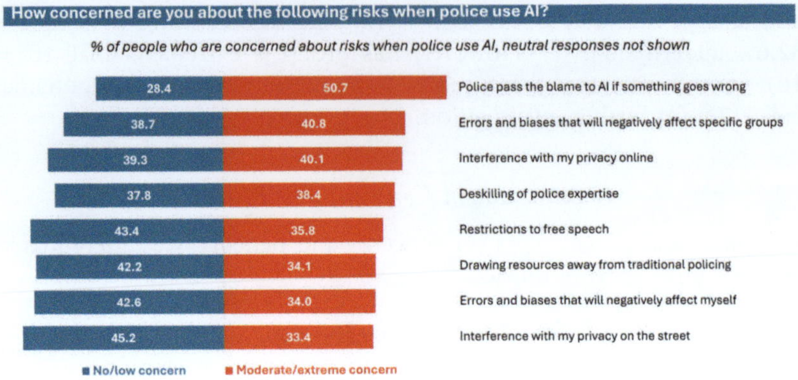


**Fig. 10.2** Public concerns against police use of AI (adapted from: Bayerl et al. [1])

of participants indicated that AI makes these aspects better, compared to 12.0% and 11.3%, respectively, who indicated that AI would make them worse (cp. Fig. 10.3). Fewer benefits were seen with regard to the protection of their own safety and even less for the protection of fundamental rights: 47.0% and 37.2%, respectively, indicated that AI makes these aspects better, compared to 17.6% and 28.5%, who indicated that AI makes them worse. This suggests that expected AI benefits are primarily associated with core policing functions, rather than societal and personal benefits.

Similar to overall support, men were more positive across all benefit types: for instance, 28.8% of men indicated that AI makes policing efforts for national security 'much better' compared to 17.5% of women. While these are no dramatic differences, this pattern holds across all benefits investigated in the survey.

An interesting effect emerged with people who reported knowledge about AI in policing. Here, benefit expectations were considerably lower compared to people who reported no knowledge of AI in policing. For instance, people without knowledge agreed to 53.7% that AI will improve the ability to protect the safety of people they care about, compared to only 7.7% in the expert group. Again, this pattern holds across all benefits, suggesting a much more critical stance in people with policing AI expertise. This is despite the fact that the overall support for police of AI is higher in police AI experts (63.4% strongly support) than in lay people (10.4% strongly support).

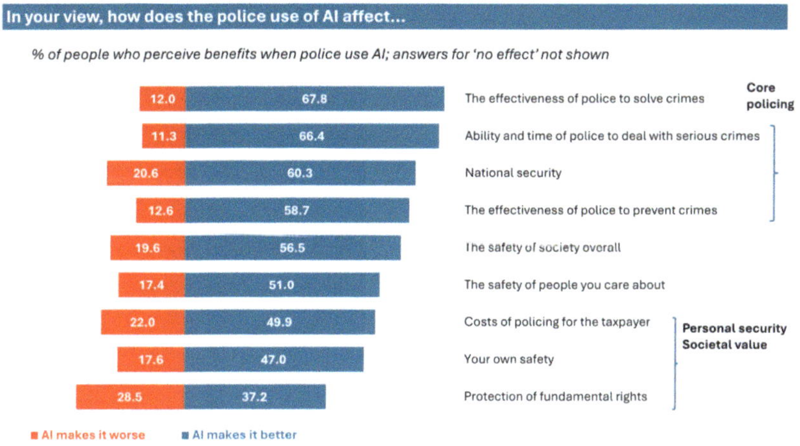


**Fig. 10.3** Benefits perceived from the use of AI by police (adapted from: Bayerl et al. [1])

This imbalance is a strong indication that how support is measured has a considerable impact on results and may tap into different aspects of public attitudes, and that overall acceptance does not guarantee acceptance for specific usages or benefits narratives.

#### *10.3.2.3 No Effect of AI for Policing*

Additional insights come from answers that claim that 'AI has no effect' on policing. More than a third of the public considered AI as being ineffective for their own safety (35.4%), the safety of people they care about (31.5%), as well as the protection of fundamental rights (31.5%). Fewer doubts existed on the impact of AI for the effectiveness of police to solve crimes (20.2% indicate 'no effect') and on national security (19.2% indicate 'no effect', cp. Fig. 10.4). Similar to overall benefits, these findings underscore that the public sees AI's main effect in core policing functions, rather than personal ones.

### *10.3.3 Public Trust in AI Use by Police and AI Decisions*

The picture on public trust is mixed. Slightly over 50% indicated that they trust the police to use AI to support their decision-making, 15.4% of these 'very much' (cp. Fig. 10.5). This still leaves 1 in 4 (25.0%) who

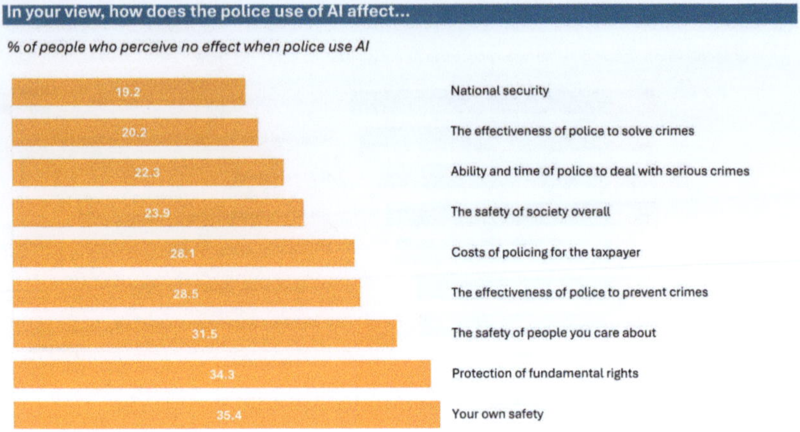


**Fig. 10.4** Perception of 'no benefits' from AI use by police across deployment areas (adapted from: Bayerl et al. [1])

distrusted AI for decision support and 1 in 5 (20.2%) who were neutral about the issue. Public trust was considerably lower when AI is allowed to make its own decisions for policing tasks: only 36% trusted AI to make its own decisions (9.4% of these 'very much'), compared to 39.7% who distrust AI-driven decisions (15.5% 'very much'; 24.3% 'neutral').

As an additional measure of trust, we further asked how willing someone would be to provide their own personal information to train AI for policing and security purposes. The question reveals a high level of unwillingness or at least uncertainty: 40% would be (very) unwilling to provide their personal information for AI training, while 29% were unsure (cp. Fig. 10.6). This leaves a very small group of 9.3% who would be 'very

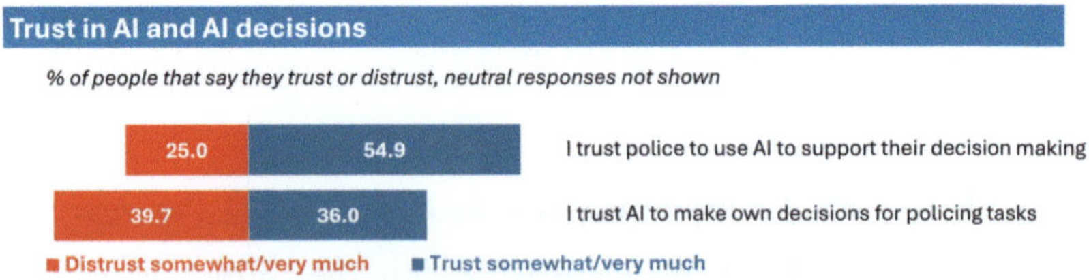


**Fig. 10.5** Trust in police using AI (adapted from: Bayerl et al. [1])

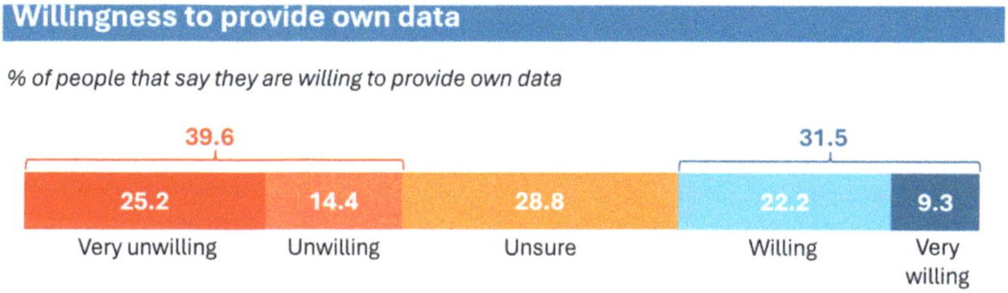


**Fig. 10.6** Willingness to provide personal data for training of policing AI systems (adapted from: Bayerl et al. [1])

willing' to provide their information. Working in a security-related profession increased the willingness to provide data to some extent: 43.9% in this group would be willing to provide their own data for the training of police AI (15.5% of these very willing), compared to 31.3% who are not linked to a security-related profession (with 9.2% of these very willing).

### *10.3.4 Public Reactions to Different AI Deployment Scenarios*

As the section on specific benefits demonstrates (Sect. 10.3.2.2), public attitudes are impacted by the specific purpose for which the police deploy their AI capabilities. As part of the consultation, we therefore included 32 disparate AI deployment scenarios to map out systematic variations in public reactions.

The 32 scenarios covered existing as well as possible (future) AI uses in policing and security. The purposes varied with respect to their deployment area (e.g. migration, child sexual exploitation, counter-terrorism, burglary, forensics, etc.), the deploying organisation (police, UK Home Office, UK Border Forces), their purpose (crime prevention, crime investigation including the identification of perpetrators, internal police efficiencies, etc.) and the risk level for privacy intrusions (low vs. high). The full list of AI deployment scenarios is provided in Table 10.2.

For each deployment scenario we assessed:

- its general acceptability,
- the degree to which AI is trusted in this context,
- the expected level of monitoring by an independent oversight body.

**Table 10.2** List of AI deployment scenarios

| |
|---|
| **Scenario 1**: Reviewing evidence and producing a case file are traditionally done manually by police officers. Your local police force uses an AI tool to automatically review the evidence and to produce a case file summary. The AI-generated case summaries are used by solicitors to decide whether to charge someone with a crime. |
| **Scenario 2**: To prevent small boat crossings from France to UK, UK border forces use AI sensors to flag up potential crossings in France. The sensors automatically notify UK border forces to intercept the boats and prevent them from landing in the UK. |
| **Scenario 3**: UK border police uses AI to investigate the online footprint of all foreign citizens who come through British borders. The automatic scan flags up any indications of criminal/terrorist activities or communication online to identify potential hostile intent against UK and impact decisions whether a person can enter the country. |
| **Scenario 4**: Your local police force uses an AI-driven robot community policing officer to autonomously patrol your neighbourhood and areas associated with high crime. If the robot identifies a potentially critical situation, it will contact the closest police station or other emergency services. |
| **Scenario 5**: Your local police is aware that a child is at risk of sexual exploitation. A fast-food restaurant handed over CCTV pictures of a meeting between the child and an unknown adult. The police places smart cameras in the area that scans people's faces to find the adult involved. |
| **Scenario 6**: Investigating child sexual abuse material causes psychological stress and mental health issues in many officers. Your local police force deploys an AI tool to scan images/videos of child sexual abuse. The AI tool automatically assigns grades of severity of the abuse safeguarding officers from viewing the material. |
| **Scenario 7**: AI tools are increasingly used to generate large quantities of child sexual abuse material. This poses a problem for police, as they do not know whether the children, they are trying to identify, are real or AI generated. Your local police introduce an AI tool to identify whether an image/video is AI generated, so police can focus their efforts on helping real victims. |
| **Scenario 8**: Your local police force uses an AI tool that scans social media for signs of impending terrorist attacks in your city. The AI creates alerts for police to start direct surveillance of suspects, or in case of an immediate threat, police receive an automatic request to arrest them. |
| **Scenario 9**: Your local police force uses an AI tool that scans social media for signs of serious and organised crime. This includes people advertising drugs for sale, evidence of firearms use and human trafficking. AI helps the police to identify the perpetrators and to safeguard victims. |
| **Scenario 10**: An airport uses an AI trained on gait analysis to identify unusual behaviours, such as nervousness, hesitation or erratic movements that could indicate terrorist intent. Any warning markers are sent to airport security and police to initiate a response. |
| **Scenario 11**: Your local police force is investigating a burglary. Very little evidence has been recovered, but forensics officers identified a partial fingerprint. The police force uses AI to enhance the print to then search for matches in a national fingerprint database. |

(continued)

**Table 10.2** (continued)

| |
|---|
| **Scenario 12**: a police force uses an AI tool to analyse DNA samples from cold cases. The AI tool uses predictive analysis to reconstruct missing sequences of partial DNA profiles, which the police force uses to search for matches in a national DNA database |
| **Scenario 13**: Your local police force uses an AI tool to speed up their investigations into digital devices such as mobile phones. When a phone is seized from a suspect, the AI searches the phone to quickly assess location data, images, records of phone calls and the content of text communication that may be relevant to the case |
| **Scenario 14**: Your local police force uses an AI tool to scan online profiles of local groups for evidence of hate speech (e.g. violence against vulnerable communities or people based on race, ethnicity, gender, religion or sexual orientation). The AI identifies individuals that seem to post hate speech and flags them for further investigation |
| **Scenario 15**: AI-generated content of politicians, actors and other celebrities (deepfakes) are being used to spread false information and encouraging public disorder or hate. Your local police force uses AI investigation tools to automatically identify the creators of AI-generated deepfake videos to prevent further spread of disinformation |
| **Scenario 16**: The courts use AI to speed up the justice process. The AI decides whether charged suspects should be bailed by the court or be remanded in prison and also calculates sentences for those found guilty |
| **Scenario 17**: The Home Office deploys an AI to speed up decisions regarding early release of prisoners. The AI reviews applications made by prisoners and decides, whether they can serve the remainder of their sentence in the community |
| **Scenario 18**: The Home Office uses an AI-based visa and immigration system to make decisions more quickly. The AI automatically assesses visa and settlement applications and decides on rejection of unsuitable applicants |
| **Scenario 19**: UK security services use AI to detect insider threats to the security of national infrastructure. They specifically aim to protect water and electricity supply as well as nuclear reactions from foreign hostile actors |
| **Scenario 20**: Your local police force uses an AI system to automate the purchase of uniforms and similar items to free up police time. The AI system identifies low stock levels, assesses the suitability of an item offered for sale and directly submits a purchase order to the company |
| **Scenario 21**: Before a demonstration, your local police use AI capabilities to monitor social media feeds for signs of potential intent of violence or crimes. People who are flagged up are prevented from joining the protest |
| **Scenario 22**: During a demonstration, your local police use AI-based sensors to automatically identify locations where unrest may break out and lead to violence. If the sensors detect potentially problematic patterns, the AI system automatically direct police resources to these areas |
| **Scenario 23**: Your local police force uses an AI system to take emergency 999 calls. The AI interacts with the caller and decides whether and when to dispatch response officers |

(continued)

**Table 10.2** (continued)

| |
|---|
| **Scenario 24**: a case of shoplifting at a large supermarket in your area has been reported to the police. The stolen items are worth just under £100. The supermarket has handed over pictures of its surveillance camera to your local police force. The police run AI-based facial recognition on the pictures to identify potential suspects |
| **Scenario 25**: a disused warehouse in your area is suspected to be used to store both weapons and drugs for an organised crime group. A nearby surveillance camera has captured a facial image of an individual frequently leaving with packages. The police set up an AI-based live facial recognition system to identify the person in real time |
| **Scenario 26**: When on the beat, officers in your local police force use smart glasses that access information about people on the street in real time. If the AI flags up concerns about a person, the officers initiate a stop-and-search |
| **Scenario 27**: Your local police force uses an AI system to take non-emergency 101 calls. The AI interacts with the caller and provides advice to their queries |
| **Scenario 28**: Police training programmes in your local police utilise AI to generate realistic training scenarios. Police students learn through interaction with highly convincing AI-controlled virtual witnesses, victims, or suspects |
| **Scenario 29**: Your local police force uses an AI-assisted system while interviewing suspects. The AI analyses words, micro-expressions and tone of voice of suspects to assess credibility and to provide investigators with guidance for questioning |
| **Scenario 30**: Your local police force uses an AI system to scan social media profiles of police officers. The AI automatically flags up officers potentially acting disproportionately or exhibiting racist, discriminatory, or otherwise problematic behaviour |
| **Scenario 31**: Your local police force uses AI to analyse police records, social media and publicly available data for indications of domestic violence to recommend preventive interventions for potential victims |
| **Scenario 32**: Your local police force deploys automated, AI-controlled police vehicles that autonomously patrol areas and independently identify and report suspicious activities, traffic violations and accidents |

#### *10.3.4.1 Acceptance Levels across Usage Situations*

The comparison across scenarios demonstrates how strongly public reactions are influenced by the type of AI deployment situation. The lowest levels of acceptance were found for AI-automated 999 emergency calls (scenario 23) and automated decisions about the early release of criminal suspects and prisoners (scenario 16) with only 27.4% and 29.3% in the sample indicating some or strong acceptance. In contrast, the highest levels of acceptance were observed for AI use to identify (potential) perpetrators of child sexual exploitation (62.4%; scenario 5), perpetrators of shoplifting in a neighbourhood store (58.7%; scenario 24) and (potential) terrorists crossing UK borders (58.0%; scenario 3). These findings suggest that the identification of potential suspects is one of the purposes where AI finds most acceptance.

#### *10.3.4.2 Trust Levels across Usage Situations*

Considerable variation in opinions can also be observed for the level of trust in AI use by policing. Again, trust was lowest where AI may cause direct harms to individuals themselves, for instance, where AI replaces 999 dispatchers (26.3%; scenario 23) or where AI makes decisions about an early release of prisoners (26.6%; scenario 16). Scenarios with low trust and low acceptance levels were thus closely linked.

In contrast, the highest degree of public trust emerged for using AI to improve the administrative efficiency of police forces by automating the purchase of police uniforms (54.5%; scenario 20) but also again for the protection of a child from sexual exploitation (53.2%; scenario 5). This seems to indicate that trust in police use of AI is highest both for situations in which AI deployments carry low risks for the public and for situations in which AI can reduce serious harms to vulnerable groups. It is worth noting, however, that even the highest trust levels are only slightly above 50%, indicating a limited degree of trust in AI use by police overall for specific usage scenarios.

#### *10.3.4.3 Need for Independent Oversight*

The context also affects the degree to which the public expects monitoring by an independent body for each AI deployment. The strongest need for independent oversight emerged for scenarios in which AI makes its own decisions in high-risk situations (e.g. when AI is used to automatically review evidence (63.7%; scenario 1) or to automatically identify AI-generated child sexual exploitation material (62.1%; scenario 7)). The lowest need for independent oversight was found for the purchase of uniforms (36.7%; scenario 20) and the deployment of AI sensors during demonstrations to direct police resources (45.2%; scenario 22).

#### *10.3.4.4 Links Between Acceptance, Trust and Independent Oversight*

When comparing acceptance, trust and need for oversight, it is clear that oversight by independent bodies is a broadly shared concern, independent of acceptance and trust.

As may be expected, results show strong expectations for independent oversight in situations with low trust and acceptance (e.g. scenario 23 in which an AI system takes emergency 999 calls or scenario 17 in which AI is deployed to speed up decisions regarding the early release of prisoners). However, independent oversight is also expected in scenarios that show considerable levels of trust and acceptance (e.g. scenario 7 in which an AI tool identifies whether a CSAM image/video is AI generated). Monitoring by an independent oversight body is thus not only needed in situations of low public acceptance or trust but expected by the public whenever police deploy AI.

### 10.3.5 *The Role of AI Accountability for Public Trust*

The importance of independent oversight together with concerns about negative AI consequences and of police 'passing blame to AI if something goes wrong' points to questions of accountability and trust in policing organisations.

Asked directly, how much they would trust police to use AI accountably, slightly over 50% indicated they would (16.3% of these very much). This leaves 1 in 5 (21.1%) who were neutral on the issue and 1 in 4 (25.6%) who indicated distrust in police to use AI accountably.

To get a better understanding of how to improve trust in AI policing, we provided five measures that are commonly proposed as safeguards (see Fig. 10.7). *Explainability* and *justification* emerged as the least effective measure from the public's perspective: 39.0% of respondents signalled that this would give them only slight trust or no trust at all (cp. Fig. 10.7). This is higher than the 30.4% who indicated that this measure would give them considerable or even very high trust. *Compensation and redress* fared slightly better (33.3% considerable/very high trust vs. 35.5% limited/no trust), as did enforcement of *changes to the AI system* in cases where the police use of AI leads to negative consequences (33.4% considerable/very high trust vs. 35.2% limited/no trust). The most effective measures, and the only ones where 'trust' percentages were higher than those for 'no trust', were the close monitoring of AI use and effective accountability (38.5% and 40.0%, respectively). Thus, none of the five common trust measures may be able to instil trust by themselves; however, the results do give indications to public preferences for assurance measures.

In this context it is compelling that 75.5% of the public wanted a mandatory accountability process before the police deploy AI. However, less than 20% of the public thought that AI accountability in policing is sufficiently established at this moment (cp. Fig. 10.8).

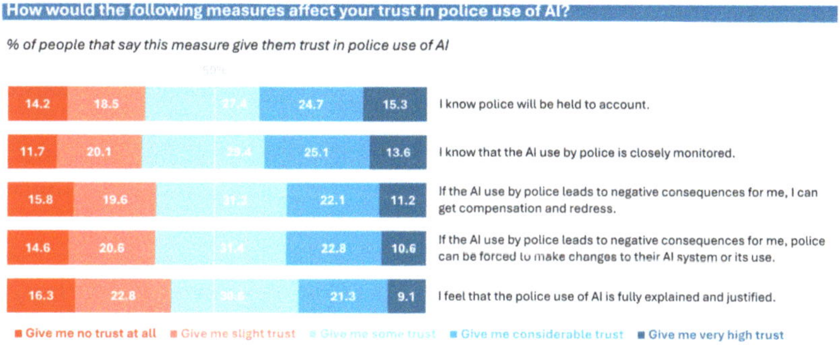


**Fig. 10.7** Degree of trust due to different accountability measures (adapted from: Bayerl et al. [1])

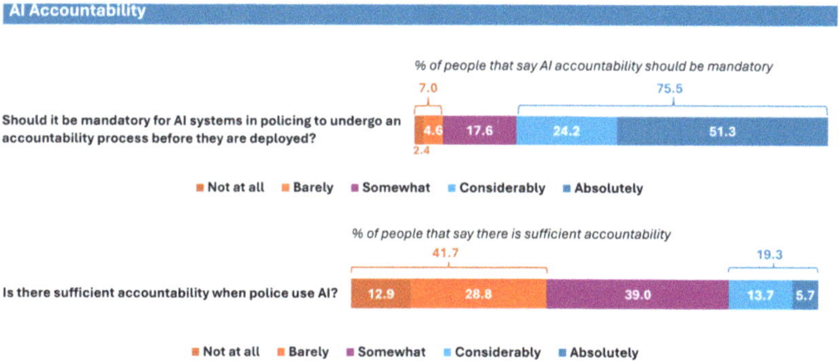


**Fig. 10.8** Perceptions of current and required AI accountability (adapted from: Bayerl et al. [1])

## 10.4 Implications for AI Deployments

Results overall show that a majority of the UK public supports AI in policing, particularly where it enhances the effectiveness of policing and addresses serious harms such as child sexual exploitation and terrorism. In the same regard, findings suggest that AI should be used to support (human) decision-making, as trust drops sharply when AI is expected to operate autonomously in making policing decisions. This may be explained by the fact public concerns centre around accountability gaps, such as a fear that police 'will blame AI' if decisions go wrong and the view that accountability is currently (somewhat to totally) lacking.

Accordingly, there is a widely shared expectation of independent oversight, even for deployments with high levels of acceptance. Independent oversight thus must be universal—with an expectation of mandatory accountability checks before AI deployments.

The consultation thus gives important guidance on citizen expectations on how to approach and implement AI use in policing. Policing should, as a minimum:

- Establish statutory, independent oversight and accountability mechanisms.
- Enforce mandatory pre-deployment reviews and bias testing.
- Strengthen transparency and citizen engagement to maintain legitimacy.
- Apply a tiered risk-based deployment model, prioritising crime-solving and child protection, while restricting high-risk uses such as AI-led 999 triage and prisoner release.

Together, these measures will create an 'accountability-first' AI model, which can harness AI's benefits for policing while supporting public acceptance and trust. The following section provides detailed recommendations on how to achieve accountable AI deployments in policing in line with public expectations.

### *10.4.1 Recommendations for AI Deployments in Policing to Ensure AI Accountability and Support Public Trust*

#### *10.4.1.1 Measures for Governance and Oversight*

- Create a statutory AI Oversight Authority for policing and security that is independent from policing and law enforcement, with powers to monitor, audit and sanction the improper use of AI.
- Mandate pre-deployment accountability reviews for all AI systems, including tests for fairness, reliability and proportionality.
- Require regular post-deployment audits and publish findings to appropriate authorities and oversight bodies.
- Build cross-institutional coordination to avoid siloed regulation.

#### 10.4.1.2 *Measures for Enshrining Accountability and Responsibility*

- Introduce a national framework, modelled on Ombudsman-style approaches, that regulates redress and compensation when AI errors cause errors or harm.
- Require record-keeping ('AI decision logs') for all uses in critical policing tasks to ensure the traceability for courts, oversight bodies and citizens.
- Caution against the transfer of 'blame' from police to AI systems; ensure accountability always rests with the provider and/or user deploying the AI.

#### 10.4.1.3 *Measures to Improve Transparency and Public Communication*

- Implement Transparency-by-Default requiring police forces to disclose where, why and how AI is being deployed, including details of training data sources (under consideration of operation/legal restrictions such as national security or victim protection).
- Produce plain-language impact assessments for each AI deployment that are available to the public (under consideration of operation/legal restrictions).
- Ensure public awareness of AI deployments by clarifying benefits, risks and limits of AI in policing to counter misinformation.

#### 10.4.1.4 *Measures for Deployment and Risk Management*

- Implement a tiered deployment model:
  - Light oversight for low-risk efficiency AI (e.g. logistics, scheduling, admin).
  - Monitoring with human-in-the-loop for medium-risk operational AI (e.g. crime pattern analysis, forensics support, life facial recognition).
  - Tight restriction or prohibition for high-risk decision AI (e.g. sentencing, prisoner release, emergency call triage).
- Require final human decision-making for any system that affects liberty, safety, rights (e.g. detention, bail, surveillance) or complaints.

- Review AI uses with low public acceptance and high risk (e.g. AI replacing 999 operators, autonomous patrol robots) to ensure benefits outweigh the risks.

#### 10.4.1.5 *Measures to Remove/Mitigate Biases and Ensure Fairness and Equality*

- Require bias testing across demographic groups before AI procurement and during AI deployment.
- Establish documentation of any known biases and, where required, ensure appropriate mitigation measures.
- Establish citizen-informed fairness benchmarks, especially where vulnerable or historically over-policed groups are involved.
- Ensure the independent testing of systems before approval.
- Mandate regular Equality Impact Assessments for AI tools in use.

#### 10.4.1.6 *Measure for Privacy and Data Protection*

- Introduce a 'necessity and proportionality test' before deploying AI with large-scale data collection from the public (e.g. facial recognition, social media monitoring).
- Require data minimisation and purpose limitation in AI deployments, aligned with GDPR and relevant national regulations.

#### 10.4.1.7 *Measure for Improved Citizen Engagement and Trust-Building*

- Establish citizen panels or advisory councils on AI policing to provide continuous input into governance.
- Institutionalise regular citizen consultations at national and regional levels to capture evolving public sentiment.
- Ensure citizen consultations comprise a comprehensive spectrum of citizen demographics and capture subgroup-specific benefit expectations, as well as subgroup-specific concerns.
- Incorporate youth perspectives into AI accountability debates, recognising generational differences in trust.

#### *10.4.1.8 Measures for Capacity Building and Training*

- Provide specialist AI literacy training for officers and staff to ensure they understand system limits and ethical considerations.
- Develop cross-disciplinary training for oversight bodies (law, ethics, data science, human rights) with a specific focus on AI accountability.
- Build a central AI in Policing Knowledge Hub for sharing best practices, red flags and audit results across police forces and connected stakeholders.

**Acknowledgements** Work in this chapter was partially funded under the AIPAS project ('AI Accountability for Policing and Security'; UKRI Engineering and Physical Sciences Research Council, Responsible AI (RAI) IA091 Grant Ref: EP/Y009800/1).

## References

1. P.S. Bayerl, B. Akhgar, F. Sampson, M. Snowden, C. Rowley, A. Paradise, Spencer, C, Citizen Consultation on the Police Use of Artificial Intelligence (AI) in the UK: Acceptance, Trust and Accountability. AIPAS Report (2025). Document available online: https://stwebsitesprodeu.blob.core.windows.net/centric-website/CENTRIC_AI+Policing_National-Citizen-Consultation (public-version).Oct2025-WEB.pdf
2. Y. Ezzeddine, P.S. Bayerl, H. Gibson, Safety, privacy, or both: Evaluating citizens' perspectives around artificial intelligence use by police forces. Polic. Soc. **33**(7), 861–876 (2023). https://doi.org/10.1080/10439463.2023.2211813
3. W. Heaven, Predictive Policing Algorithms Are Racist. They need to be dismantled (2020). Retrieved from MIT Technology Review: www.technologyreview.com/2020/07/17/1005396/predictive-policing-algorithms-racist-dismantled-machine-learning-bias-criminal-justice
4. L. Horvath, O. James, S. Banducci, A. Beduschi, Citizens' acceptance of artificial intelligence in public services: evidence from a conjoint experiment about processing permit applications. Gov. Inf. Q. **40**(4) (2023). https://doi.org/10.1016/j.giq.2023.101876
5. T.W. Hung, C.P. Yen, On the person-based predictive policing of AI. Ethics Inf. Technol. **23**, 165–176 (2021)
6. R. Pastor, F. Mignet, T. Mattes, A. Gurzawska, H. Nitsch, D. Wright, COPKIT: technology and knowledge for early warning/early action-led

policing in fighting organised crime and terrorism, in *Technology Development for Security Practitioners. Security Informatics and Law Enforcement*, ed. by B. Akhgar, D. Kavallieros, E. Sdongos (Springer, Cham, 2021), pp. 121–133
7. J. Seaman, Cyber threat prediction and modelling, in *Artificial Intelligence and National Security*, ed. by R. Montasari (Springer, Cham, 2022), pp. 113–156

**Petra Saskia Bayerl** is Professor of Digital Communication and Security and Head of Research at CENTRIC. Her research addresses societal implications and governance of new technologies in security and policing. She has been co-lead for both the AP4AI and AIPAS projects and lead for national and EU-projects to improve the protection against online harms.

CHAPTER 11

# Approaches to AI Accountability and AI Compliance-by-Design for R&I Projects in Policing and Security: Lessons Learned from STARLIGHT

*Christopher Guiver and Helen Gibson*

## 11.1 Introduction

As Artificial Intelligence (AI) technologies have developed over the past 10 years, the need for governance—legislatively, organisationally, ethically and morally—has emerged. This has led national and supranational bodies to adopt differing approaches to governance; the European Union, for example, through the AI Act,[1] has adopted an approach grounded in

C. Guiver · H. Gibson (✉)
CENTRIC (Centre of Excellence in Terrorism, Resilience, Intelligence and Organised Crime Research), Sheffield, UK
e-mail: h.gibson@shu.ac.uk

[1] Regulation (EU) 2024/1689 of the European Parliament and of the Council of 13 June 2024 laying down harmonised rules on artificial intelligence and amending Regulations (EC) No 300/2008, (EU) No 167/2013, (EU) No 168/2013, (EU) 2018/858, (EU) 2018/1139 and (EU) 2019/2144 and Directives 2014/90/EU, (EU) 2016/797



B. Akhgar and P. S. Bayerl (eds.), *Accountability for Artificial Intelligence in Policing and Security*, Security Informatics and Law Enforcement, https://doi.org/10.1007/978-3-032-41284-3_11

primary legislation, albeit supported by extensive guidelines and underpinned by long-standing ethical input [4]. Other jurisdictions such as the UK, at least for now, have a general framework, while devolving implementation rules to sector-specific regulators and bodies [7].

In this chapter, we focus on how the EU's legislation-first approach impacts the development of policing and security technologies throughout the innovation and development cycle. Section 11.2 introduces the positioning of AI development within the EU's Research and Innovation framework, the impact the AI Act has had on this development to date and the challenges and available mechanisms for addressing them. Section 11.3 follows by briefly describing the AP4AI (Accountability Principles for AI) project, its relationship and differences to AIPAS and the specific elements adapted to the EU context. Section 11.4 introduces the STARLIGHT project: a large R&I project designed to develop AI technologies for security and elaborates it as an example for implementing AI accountability and compliance-by-design in research. The final section reflects on the lessons learned and proposes introducing a three-phase approach to utilising AP4AI in future EU R&I projects for policing and security.

## 11.2 AI in Security Research

### *11.2.1 The Role of Research and Innovation Projects in the AI Development Cycle*

Research and innovation (R&I) projects funded through European programmes such as Horizon Europe play a crucial role in the development lifecycle of AI technologies for security applications in Europe. These projects serve as experimental environments where new analytical techniques, technological tools and operational concepts can be developed and tested before potential adoption by law enforcement and security organisations.

Unlike commercial technology development, security-focused R&I projects typically involve multi-stakeholder consortia composed of research institutions, technology developers, law enforcement agencies and policy experts. This collaborative structure enables interdisciplinary

and (EU) 2020/1828 (Artificial Intelligence Act) (Text with EEA relevance) OJ L, 2024/1689 (2024).

innovation while simultaneously introducing governance complexities, particularly when projects aim to develop technologies that may ultimately be deployed into operational policing environments.

In particular, such projects pose specific legal and ethical challenges: they must ensure compliance with research ethics requirements while also positioning themselves in a way that could enable future deployment in an operational environment. This is abundantly clear when we consider the axis between AI and data protection. For example, a core requirement for many police investigations is to identify the perpetrator and/or the victim. This necessitates the involvement of personal data, with clear and justified reasons under the Law Enforcement Directive (LED).[2] In research, processing personal data in a similar manner may not be necessary and proportionate, nor may it comply with other principles under the General Data Protection Regulation (GDPR), for the purposes of carrying out the research. This tension can affect the development and evaluation of such tools; thus, AI researchers need a good understanding of the interplay between these different legal frameworks in the operational environment, as well as in the research one. Consequently, governance mechanisms for accountability, transparency and regulatory compliance must be integrated into technical development processes at an early stage. In parallel, challenges with data governance through AI training must also be addressed, particularly issues of consent, discrimination, fairness, ownership, privacy, surveillance and trust [6].

One such method of addressing this is through embedding governance and accountability approaches upfront in the project design. Typically, AI systems developed within such projects often evolve iteratively throughout the project lifecycle. Therefore, ensuring such checks and balances are embedded into the project, at the correct intervention points, is crucial.

[2] Directive (EU) 2016/680 of the European Parliament and of the Council of 27 April 2016 on the protection of natural persons with regard to the processing of personal data by competent authorities for the purposes of the prevention, investigation, detection or prosecution of criminal offences or the execution of criminal penalties and on the free movement of such data and repealing Council Framework Decision 2008/977/JHA. OJ L 119 (2016).

## 11.3 STARLIGHT Project

In this chapter, we use the Sustainable Autonomy and Resilience for LEAs using AI against High priority Threats (STARLIGHT)[3] project as an illustrative example of different approaches to embedding legal and accountability requirements into EU R&I. STARLIGHT aimed to bring law enforcement agencies (LEAs), researchers, industry and practitioners in the security ecosystem together under a coordinated and strategic effort to innovate and develop AI solutions from research and increase their maturity towards a level where they could be realised in operational practice. The project engaged 50 partners directly, of which 13 were law enforcement agencies.

STARLIGHT was an excellent test case as the variety of AI solutions covered a number of different stages of the investigation process—from dataset acquisition and development to data (pre-processing), AI-based analysis and analytics, as well as offensive and defensive cybersecurity solutions. This led to a wide-ranging AI toolkit that encompasses tools designed to redefine how we approach security challenges with modern, innovative technology. These tools exemplify AI's potential to tackle complex security issues and include visual and audio/text processing analytics designed to empower law enforcement agencies with advanced capabilities for enhanced security and investigation. (For more details on the outcomes of STARLIGHT, see the STARLIGHT Project Results Brochure[4]).

### 11.3.1 *The Impact of the EU AI Act on Security Research*

The development of the STARLIGHT project occurred during a period of significant regulatory change within the European Union. The proposal and subsequent adoption of the EU Artificial Intelligence Act introduced a comprehensive risk-based regulatory framework to ensure that AI systems placed on the European market meet requirements for safety, transparency, accountability and the protection of fundamental rights.

[3] https://starlight-h2020.eu.

[4] STARLIGHT Project Results Brochure: https://starlight-h2020.eu/news/2026-01/starlight-project-results-brochure.

The AI Act utilises a risk-based approach to compliance with 'riskier' applications (labelling them as 'high-risk') and requiring stricter governance processes (see Chap. 4). In particular, certain AI applications used in policing and security contexts may fall within the category of high-risk AI systems, which are subject to enhanced obligations concerning documentation, risk management, human oversight and traceability. Although research projects may not always produce immediately deployable systems, technologies developed within such projects frequently serve as precursors to operational tools. As a result, research projects must increasingly anticipate future regulatory requirements and integrate governance mechanisms to support compliance as systems transition from experimental environments to operational use.

### *11.3.2 The Role of AI Accountability in R&I Projects*

As we have seen in the other chapters (e.g. Chap. 2 for accountability, Chap. 6 for organisational best practices and Chap. 10 for societal expectations), policing and security organisations do not just have legal obligations; they are also accountable to the public they serve. Whether that is in the UK or Europe, the functioning of a democratic society relies on trust and legitimacy of the police. Ensuring accountability in the development of AI systems intended for policing and security applications, where the potential for catastrophic consequences and serious reputational damage is high, is therefore essential. The documentation of governance mechanisms such as risk assessments, development records and auditability processes can also serve as an important precursor to formal regulatory compliance.

Therefore, in addition to the ethics and legal compliance research projects need to consider, ensuring that the AI systems developed demonstrate that they align with the moral obligations required by policing is also imperative. Establishing such governance practices during the research phase can significantly reduce compliance risks in later stages of technological deployment, particularly when aligned with the policing organisations that will ultimately use them. This chapter therefore explores how accountability frameworks can be implemented within research environments to support AI compliance-by-design, drawing on the practical experience of the projects.

## 11.4 AP4AI: An Accountability Framework for AI in Policing and Security

The Accountability Principles for Artificial Intelligence (AP4AI)[5] framework was developed to provide structured guidance for the responsible development and deployment of AI systems in the policing and internal security domains, with a focus on the European context [1].

The framework emerged from collaborative work involving law enforcement organisations, academic researchers, policy experts and technology specialists. Its primary objective was to establish governance mechanisms capable of supporting lawful and responsible AI adoption and development within the security and policing domain while maintaining public trust in law enforcement institutions [1, 2].

As with AIPAS (Chap. 5), AP4AI translates the aforementioned accountability principles into practical assessment mechanisms that organisations can use to evaluate the governance and compliance status of AI systems throughout the development lifecycle [2].

Furthermore, during implementation of the framework in the STARLIGHT project, AP4AI underwent significant development to align with the EU AI Act. This provided a unique selling point for AP4AI, providing one of the first self-assessment approaches to AI Act compliance, while maintaining the crucial link to accountability mechanisms. Furthermore, neither AI, policing nor accountability can be considered a static process. Therefore, as research into AI accountability in policing evolved under AIPAS (Chap. 5), a harmonisation to a single set of principles was also streamlined into AP4AI, ensuring that the AP4AI framework continued to adopt best and accepted practices in AI accountability for policing and security.

To operationalise the AP4AI framework, a web-based self-assessment platform was deployed to support developers, researchers and practitioners in evaluating whether their AI systems met accountability and regulatory requirements. Practically speaking, the AP4AI self-assessment platform is aligned to the features and interface of the AIPAS platform (Chap. 9); thus, we only describe the elements unique and relevant to AP4AI here.

In the configuration used during the STARLIGHT project, the AP4AI assessment consisted of three main sections:

5 https://ap4ai.eu.

1. **General System Information**: this section collects contextual information about both the AI system under assessment and the organisation conducting the AP4AI evaluation. It seeks to establish whether the system falls within the scope of the EU AI Act, including determining whether it may be classified as a prohibited AI practice or a high-risk application. In addition, the section gathers further details regarding the characteristics of the AI system itself, including whether the technology has been developed internally or procured as third-party software.
2. **Accountability Principles Assessment (AP4AI)**: the AP4AI section enables practitioners within the security and justice domains to assess, review and document the accountability of their AI applications throughout the entire system lifecycle. The framework provides a comprehensive approach to AI accountability by addressing governance considerations across the design, procurement and deployment of AI tools, as well as their operational outcomes and societal impacts. Through this structured assessment process, practitioners are able to harness the benefits of AI technologies while demonstrating responsible and transparent use to oversight bodies and the wider public. The assessment of accountability is split between the eight separate sections, with direct links to the 12 accountability principles.
3. **Alignment with EU AI Act Requirements**: this section enables security and justice practitioners to assess, review and document the alignment of their AI applications with the risk-based regulatory requirements set out in the EU AI Act. The assessment framework incorporates the obligations defined for AI providers, deployers, importers, distributors and authorised representatives under Regulation (EU) 2024/1689. These requirements are structured within the platform to enable users to systematically assess their preparedness for compliance and identify potential gaps in governance, documentation or operational practices before deploying AI systems. Each subsection of this section of the assessment directly targets specific articles of the AI Act, with an explicit focus on the compliance obligations for high-risk AI system, specifically:

(a) **Risk Management System** (Article 9): this ensures a holistic approach to AI safety and reliability, considering risks for individuals and society. Questions explicitly consider the user context, such as the intended users' expertise, training and environment.
(b) **Data and Data Governance** (Article 10): for both development of AI models, testing and training and deployment processing.
(c) **Technical Documentation** (Article 11) and **Record-Keeping** (Article 12): address the keeping of the output and outcomes of the AI system in logs, as well as traceability, risk monitoring and technical documentation, alongside conformity with regulatory standards.
(d) **Transparency and Provision of Information** (Article 13): to ensure relevant stakeholders can interpret outputs appropriately.
(e) **Human Oversight** (Article 14): high-risk AI systems should be designed and developed in such a way that natural persons can oversee their functioning, ensure that they are used as intended, and that their impacts are addressed over the system's lifecycle.
(f) **Accuracy, Robustness and Cyber Security** (Article 15): high-risk AI systems shall be designed and developed in such a way that they achieve an appropriate level of accuracy, robustness and cybersecurity, and that they perform consistently in those respects throughout their lifecycle.
(g) **Obligations for providers, deployers, importers, distributors and authorised representatives** of high-risk systems and other parties (Sect. 11.3).
(h) **General Purpose AI:** the AI Act sets forth specific requirements for AI systems that can be considered as general purpose AI (GPAI).
(i) **Fundamental Rights Impact Assessment** (FRIA): a mandatory pre-deployment review by deployers of the impact on fundamental rights of the AI system.

The complete assessment contains approximately 178 questions, depending on the type and functions of the AI system being assessed, addressing issues including system design, data governance, documentation practices, risk management procedures and oversight mechanisms.

Most questions are structured using a three-point response scale, resulting in the contribution of the following analytics in the final report of 'achieved', 'in progress' or 'needs addressing', accompanied by free-text fields that allow respondents to provide explanations and references to supporting documentation.

At the end of an assessment, reports are generated that provide an overview of the overall assessment, areas requiring remedial compliance actions and a detailed overview of all AI Act-related responses. These can be used as supporting documentation for compliance and accountability processes.

## 11.5 AI Accountability and Compliance in R&I: STARLIGHT as Test Case

### *11.5.1 STARLIGHT Development and Assessment Process*

STARLIGHT, as the largest EU-funded project focused on the development of AI technologies for security, had a responsibility to demonstrate accountable and compliant use of AI within the project. The project was highly operationally focused, with a strong emphasis on co-development among practitioners, technical and research partners. Furthermore, as previously mentioned, the project was active during the drafting, trilogues and, ultimately, the political agreement on the AI Act. While the majority of the AI Act, particularly the aspects relevant to STARLIGHT, did not come into force during the project's lifetime, the project had the opportunity to be a trailblazer for EU-funded civil security projects developing AI and to contribute to best practice.

STARLIGHT extensively developed AI tools, at varying levels of maturity according to the technology readiness level scale (TRL) [3], with all tools required to undergo an AP4AI assessment prior to any piloting activity, with the stipulation that all tools should follow the 'high-risk' route of assessment to demonstrate maximum compliance and accountability, in line with Article 95 of the AI Act. Throughout the project, 60 AI tools underwent an AP4AI assessment, with up to 9000 individual questions answered (Figure 11.1 shows an example from the AP4AI dashboard).

Furthermore, in STARLIGHT, organisations were not simply left to complete the assessments themselves; they were supported by structured guidance from legal and accountability experts, both within their own organisations and the projects. Each assessment received a comprehensive review from the research team, followed up through online meetings,

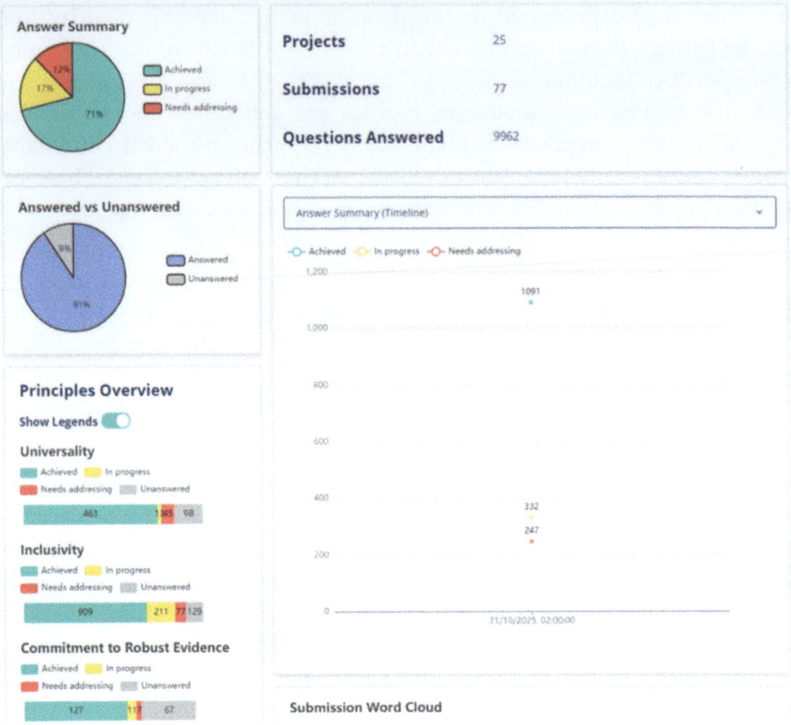


**Fig. 11.1** AP4AI assessment dashboard analytics

enabling the identification of common challenges and approaches to navigating them. Such conversations often involved addressing complex legal interpretations of how the AI Act would be applied before it entered into force. Such an approach was beneficial for both sides in understanding the complexities of the AI Act and its implementation, enabling everyone to be better prepared. Upon completion and sign-off of the assessment, tool providers were able to use their completed assessment reports to demonstrate their engagement with the compliance process.

Analytics from the platform showcased that 71% of questions answered were marked as 'achieved', 16% marked as 'in progress' and 13% marked as 'needs addressing'. This showed that, on average, assessed systems had identified themselves as meeting the majority of compliance and accountability criteria, with a smaller portion still working towards full alignment.

The platform's dashboard analytics showed that the majority of the answers focused on three core domains based on the wordcount graph: the tool, data and documentation, as shown in Fig. 11.2, demonstrating strong emphasis on the practical, data-driven and evidence-based aspects of compliance and accountability.

Some of the limitations for assessments that are implemented during R&I projects are discussed later in the section; however, some of the most common challenges around the AI Act itself proved to be the following:

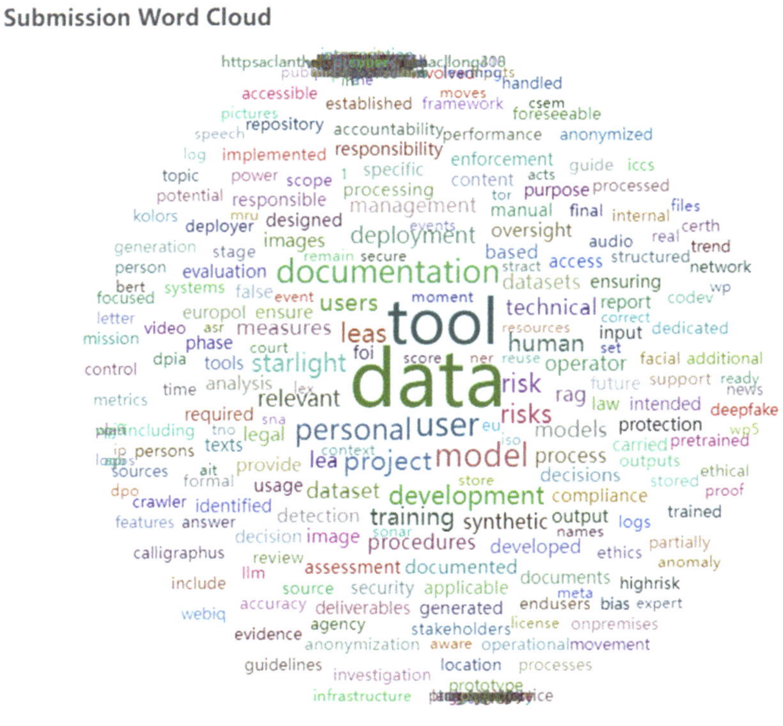


Fig. 11.2 STARLIGHT AP4AI platform wordcloud

- Classification of high-risk systems, including identifying the intended purpose of the system and the specific areas of application under Annex III.
- Ability to document whether the AI system provides an evidential function or a decision support function.
- Underlying model compliance, relating to pre-trained models and their training data, licensing conditions and trade-offs between accuracy, precision and compliance.
- Management of bias assessment and mitigation measures, especially for pre-trained models.
- Challenges in assessing AI components that form part of a larger AI system.
- Aspects where the emphasis is on deployer rather than provider compliance.

These elements highlight not only the challenges that research projects may face but also the intensity of the complexity faced by anyone deploying an AI system within the EU.

After the successful implementation in STARLIGHT, AP4AI was also deployed in another Horizon Europe project—Fight Against Large-scale Corruption and Organised Crime Networks (FALCON),[6] contributing to compliance activities related to trustworthy AI, which also included technical explainability of AI decisions and output trustworthiness. Unlike in STARLIGHT, the assessments took a more hands-off approach, allowing developers to undertake assessments with minimal guidance. This approach was intentional in order to assess the usability and intuitiveness of the AP4AI system and accountability framework.

A key objective in the continued development of the AP4AI platform has been to provide an intuitive and accessible self-assessment environment that enables organisations to independently evaluate the accountability and regulatory preparedness of their AI systems. As AP4AI and AIPAS are adopted by further projects and organisations, the need for a decentralised self-assessment process becomes paramount. Organisations must be able to evaluate their AI systems and their use thereof independently. Technical support for the platform and guidance on interpreting assessment criteria remain available as required; however, the

[6] https://www.falcon-horizon.eu/.

long-term objective is to enable organisations to embed AI accountability and compliance assessments within their own governance processes.

### 11.5.2 *Success and Challenges of Implementing AI Act Compliance in R&I Projects*

The AI Act has, undoubtedly, brought with it a significant increase in the number of governance and compliance factors that need to be considered within the AI development process. For high-risk systems, these are more complex than for similar legislation, such as the GDPR. The critical success of the implementation of AP4AI within STARLIGHT was the increase in literacy related to the AI Act and AI accountability considerations recognised by developers and practitioners alike. This level of engagement immediately translated itself into the request for implementation in the FALCON project, highlighting both the value of the tool and the process. By demystifying the requirements and compliance obligations, this helped many of those working in the area that they are already considering many of the necessary aspects (e.g. related to data, bias, transparency, human oversight and similar) during the course of their development processes, and that formally documenting much of this information allows for a significant step forward towards compliance.

Of course, during the early stages of the projects, the EU AI Act had not yet been finalised, meaning that precise compliance requirements were still subject to change; this unfortunately meant that updates to self-assessments had to be made by developers as time went on to continuously ensure compliance. The lack of stability in the AI Act implementation during the current period, such as with the digital simplification package, is still impacting on processes for managing compliance activities and will ultimately necessitate updates to the AP4AI framework and previously completed compliance assessments.

Alongside this, R&I projects are inherently experimental, with technologies evolving rapidly throughout the project lifecycle. This made it difficult to apply evolving governance frameworks designed for developing production systems; however, waiting until post-development to make a retrospective assessment also makes it difficult to re-engineer solutions if there was a gap in compliance. Therefore, it may also be beneficial for the system to highlight which aspects are non-negotiable to consider prior to development and deployment, and which need continuous monitoring during the maintenance phase.

Research itself is quite a different process from developing a commercially focused product from the outset. Naturally, research teams often prioritise the art of the possible over governance documentation, making it challenging to collect evidence required for compliance assessments. This was especially difficult during the STARLIGHT project as strict oversight of over 60 assessments was a significant burden for a small team. For many security R&I projects, AI tool development is contributing to a single overall system. In these cases, there are opportunities to standardise some responses to certain questions at a project level.

The AP4AI assessments also required complex stakeholder engagement involving multiple organisations, coordinating accountability responsibilities across different institutional departments, legal frameworks and technical practices. This meant that although the assessments were targeted towards technical developers within the project, these partners often needed to draw on expertise from ethical, legal and organisational leads within their organisation in order to answer certain questions; however, such departments did not necessarily have the expertise themselves to understand such technically complex systems, particularly at the rate that AI is evolving.

Finally, with guidelines on the EU AI Act for high-risk systems only just opened for public consultation [5], there remains a risk of ambiguity regarding precise compliance requirements. As such, an assessment editor functionality of the platform has been introduced. This allows for certain assessments within projects to be altered rapidly in response to changes to the EU AI Act or the accountability framework. Implementation of this is at the project level, meaning that previously completed assessments, based on pre-guidelines compliance, are unaffected, and processes to update assessments to current guidelines in a structured and managed approach are possible.

## 11.6 Best Practice for Future Use of AI Accountability- and Compliance-by-Design Approaches in R&I Projects

The STARLIGHT project necessitated a flexible process with ad hoc feedback being given to developers and updates to the AP4AI tool being required as and when new guidance was introduced. Even at this stage, there are still new and emerging factors to be considered. The latest guidelines, for example, provide for the consideration of how even if tools may be considered as low-risk in a standalone implementation, they can become high-risk if part of a larger system [5].

In order to manage compliance assessments going forward, a three-step process has been developed to account for the dynamic nature of security R&I projects and the need for continuous reassessment as the tools increase in maturity throughout the project's duration.

The current three-phase approach is being trialled in Identify Attributes Matrix Initiative (IAMI),[7] which is developing tools for entity identification and resolution, and Deepfake Evidence and Technology for Forensic Content Oversight and Research (DETECTOR), which is developing solutions for synthetic media detection across various modalities.

As detailed in Fig. 11.3, this approach includes the following steps:

- **The Check-In Phase**: a preliminary assessment is undertaken before any major development work starts, outlining general information about the proposed AI tool, such as the proposed models used, nature and sources of training, validation and test data to be used within the project and information about organisational processes in place, which must be implemented before major development begins of the project.
- **The Check-Up Phase**: following this, a fully completed assessment is made available before the first operational test of any tool. If an AI tool is to be tested, then realistically, it should be in a position in which all questions within the AP4AI assessment would have a robust answer along with supporting documentation.
- **The Check-Out Phase**: a final iteration of the assessment is then made available before the completion of the project and for any final test activities. This can be referenced for accountability evidence and any legal auditing purposes. In this phase, LEAs or security practitioners, who would deploy the tool, are also engaged to ensure that they are aware of the requirements on them to utilise the tool in an operational setting.

At the end of each phase, the document should be reviewed by experts, who can highlight specific actions for moving towards greater

[7] https://iami-project.eu/.

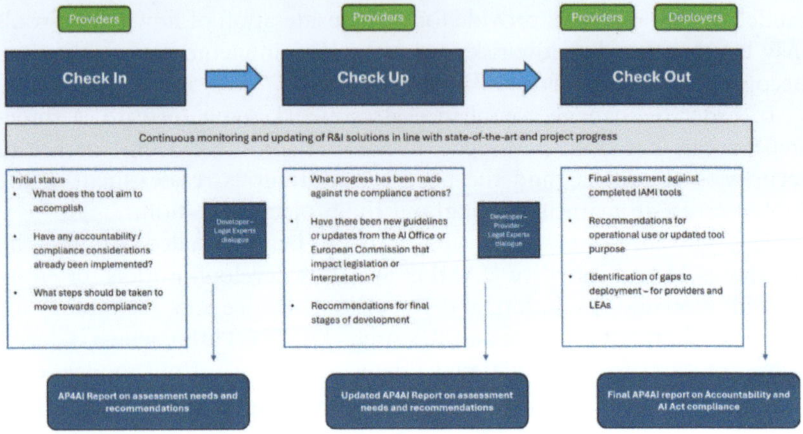


**Fig. 11.3** Three-step assessment process for AI accountability and AI Act compliance in security R&D projects

compliance, recommendations for implementation and considerations for operational use. For example, AP4AI assessments are primarily for completion by developers during the project; however, by the check-out phase, practitioners or end-users should have access to enough information and documentation, to consider how they would undertake their role as deployers of AI tools and the organisational responsibilities that accompany that.

The decision to adopt the three-stage assessment methodology was informed by evidence gathered during the implementation of the AP4AI platform in the STARLIGHT and FALCON projects. Feedback from project participants, together with the experience of facilitating the assessments and moderating the AP4AI platform, demonstrated that a single end-of-project assessment did not provide sufficient opportunity to identify and address governance and compliance issues during system development. Consequently, an iterative three-stage assessment process was introduced to embed AI accountability and compliance-by-design throughout the project lifecycle.

The methodology was designed around the following SMART objectives:

- **Specific**: establish three defined governance checkpoints (check-in, check-up and check-out) aligned with key stages of AI system development.
- **Measurable**: ensure that every AI system completes all three assessments, with each stage producing a documented AP4AI assessment report and tracking progress against previously identified recommendations.
- **Achievable**: AI systems can be used operationally by the end of the project, implementing a compliance-by-design philosophy without disrupting technical development.
- **Relevant**: evidence-based support for AI Act compliance and accountability and enabling corrective action throughout the development lifecycle.
- **Time-Bound**: complete each assessment before the corresponding project pilot or development milestone.

The introduction of the three-step approach aims to give practitioners a steady pace of assessment, grounded in accountability-by-design throughout the AI development lifecycle, rather than a box-ticking exercise at the end of the assessment. The check-in assessment establishes a baseline and highlights missing or insufficient elements of the assessment at a point where the majority of required information should be at hand. Having a designated check-in period is an anticipated action with the intention of being less disruptive to the assessment practitioner's organisation. The check-out phase evaluates the maturity of the AI system prior to operational deployment, identifies any remaining compliance gaps and produces recommendations to support deployment decisions.

Preliminary results from IAMI show that this early engagement is proving beneficial across the research and innovation teams. While most developers are easily able to define the role and purpose of the AI system, the types of data and compliance with data protection regulations, there are further challenges to improving literacy related to risk assessment, oversight, optimising stakeholder engagement and ensuring that appropriate technical documentation is in place. Addressing these aspects is a current line of research under AP4AI within these projects to ensure that the developers have the support they need to continue to effectively engage with accountability and compliance actions for research and software development.

**Acknowledgements** This research has been supported by the following funding schemes. The STARLIGHT project has received funding from the European Union's Horizon 2020 Research and Innovation Programme under Grant Agreement Number 101021797. FALCON is co-funded by the European Union under Grant Agreement Number 101121281 and by UK Research and Innovation (UKRI) under the UK government's Horizon Europe funding guarantee [grant number 10078194]. IAMI is co-funded by the European Union under Grant Agreement Number 101168272 and by UK Research and Innovation (UKRI) under the UK government's Horizon Europe funding guarantee [grant number 101168272]. DETECTOR is funded by the European Union under Grant Agreement Number 101225942.

## References

1. B. Akhgar, P.S. Bayerl, G. Mounier, R. Linden, B. Waites, AP4AI: accountability principles for artificial intelligence in the internal security domain. Eur. Law Enforcement Res. Bull. **22**(6) (2022a)
2. B. Akhgar, P.S. Bayerl, K. Bailey, R. Dennis, H. Gibson, S. Heyes, A. Lyle, A. Raven, F. Sampson, *AP4AI Framework Blueprint* (CENTRIC, 2022b)
3. European Commission, About technology readiness levels. *EURAXESS—Researchers in Motion* (2026a n. d.). Available at: https://euraxess.ec.europa.eu/career-development/researchers/manual-scientific-entrepreneurship/major-steps/trl. Accessed 15 July 2026
4. European Commission, Ethics guidelines for trustworthy AI. Independent High-Level Expert Group on Artificial Intelligence. (2019) Available at: https://digital-strategy.ec.europa.eu/en/library/ethics-guidelines-trustworthy-ai. Accessed 15 July 2026
5. European Commission, *Targeted consultation on the draft guidelines for the classification of high-risk artificial intelligence systems* (Directorate-General for Communications Networks, Content and Technology. Available at: https://digital-strategy.ec.europa.eu/en/consultations/targeted-consultation-draft-guidelines-classification-high-risk-artificial-intelligence-systems, 2026b) Accessed 15 July 2026
6. L. Floridi, M. Taddeo, What is data ethics? Philos. Transact. A Math. Phys. Eng. Sci. **374**(2083), 20160360 (2016)
7. Rough, E. (2026) AI regulation in the UK. House of Commons Library—Research Briefing. Available at: https://commonslibrary.parliament.uk/research-briefings/cbp-10003/. Accessed 15 July 2026

**Christopher Guiver** is a Researcher at CENTRIC whose work over the past three years has focused on implementing the AP4AI framework and self-assessment platform across European FCT projects. His work supports the development of AI systems that embed compliance and accountability-by-design, and that these systems demonstrate a robust regulatory compliance process throughout their development lifecycle for the EU AI Act.

**Helen Gibson** is Head of Innovation at the Centre of Excellence in Terrorism, Resilience, Intelligence and Organised Crime Research (CENTRIC). Since completing her PhD, she has more than 10 years of experience working across police and security research with recent research in AI, OSINT, cybercrime and misuse of technologies. She has extensive knowledge across the entire innovation cycle—from stakeholder engagement to conceptualisation, technology development, evaluation and legal and ethical analysis.

PART III

# Use Cases

CHAPTER 12

# AI and Biometrics in Law Enforcement in England and Wales: Governance, Legality and a Potential Way Forward

*Shazia Khan*

## 12.1 Forensic Science and Biometrics: Definitions

Forensic science uses scientific knowledge and methods to help investigate crime and provide evidence for the courts. It has strengthened criminal investigations by placing greater emphasis on physical material, scientific testing, and objective interpretation, rather than relying only on witness accounts. One important idea is Locard's Exchange Principle: when people, objects, or places come into contact, small traces may be transferred between them [56]. These traces, including biological material, can help investigators understand links between people, items, and scenes but they must be interpreted carefully and in context [45, 48].

The following definitions are used in this chapter for the various traces used in forensic analyses:

S. Khan (✉)
Metropolitan Police, London, UK
e-mail: shazia.khan@met.police.uk



B. Akhgar and P. S. Bayerl (eds.), *Accountability for Artificial Intelligence in Policing and Security*, Security Informatics and Law Enforcement, https://doi.org/10.1007/978-3-032-41284-3_12

- **Deoxyribonucleic acid (DNA)** is the hereditary material found in most human cells. In forensic science, DNA recovered from biological material such as blood, saliva, semen, hair roots, or skin cells can be analysed to produce a profile, commonly using short tandem repeat (STR) markers. DNA profiles may help identify or exclude individuals, link offences, support missing-person investigations, and review historic cases. Their value depends on reliable recovery, contamination control, interpretation and statistical reporting [10, 30, 60].
- **Fingerprints** are impressions made by friction ridge skin on the fingers, palms, toes, and soles. These ridge patterns and minutiae are generally stable throughout life and can be compared with known reference impressions. Fingerprint evidence may assist with identification, exclusion, and crime linkage but its reliability depends on mark quality, recovery method, examiner competence, verification, and careful interpretation [3, 12, 17, 54].
- **Facial comparison** is the assessment of facial images, usually comparing an unknown person in CCTV, photographs, or digital media with known reference images. It can support investigations by helping to identify or exclude a person, provide intelligence leads, or assist the court where identity is disputed. Reliability depends on image quality, pose, lighting, resolution, examiner competence, documented methods, verification, and cautious reporting [5, 21, 23, 62].

This chapter examines the role of AI for biometric processes used in the criminal justice system, including the recovery, processing, interpretation, and reporting of DNA, fingerprints and facial images. Starting with the legal and regulatory frameworks that apply to forensic science, biometric data, and AI, it proposes a practical governance framework for introducing AI into high-impact biometric processes.

## 12.2 Forensic Science Governance

In England and Wales, forensic science is supported by a statutory regulatory structure. The Forensic Science Regulator Act 2021 gives the Forensic Science Regulator legal powers, while the statutory Code of Practice (the Code) sets expectations for how forensic work should be carried out and reported. In practical terms, the Code applies across the forensic process from recovering and preserving material through to analysis, interpretation, and presentation of findings. This makes operating within a quality management system central to forensic evidence [24, 27].

Compliance or non-compliance to the Code is also reported in court. Non-compliance does not automatically create civil or criminal liability, but it may affect how forensic evidence is assessed. Poor compliance can lead to regulatory action and may give the defence grounds to challenge the reliability or weight of the evidence. Biometric evidence should therefore be supported by a clear chain of custody, documented procedures, validated methods, competent practitioners, and transparent reporting [16, 24, 27].

DNA and fingerprint recovery and processing are clearly within the established forensic governance framework. Facial identification is less directly regulated at the recovery or initial processing stage, but the Code's principles remain relevant where facial-image material is compared, interpreted or reported for evidential purposes (i.e. facial comparison).

### *12.2.1 Additional Governance for Biometrics*

Biometric material is subject to separate legal controls. The Protection of Freedoms Act (PoFA) 2012[1] was introduced after concerns about indefinite police retention of DNA and fingerprints. It sets rules on when DNA profiles and fingerprints must be deleted, when they may be kept and when longer retention may be allowed for national security reasons. This means biometric use must be assessed not only by scientific value, but also by whether retention and searching are lawful.

Later amendments shaped compliance in national security and counter-terrorism cases by extending the maximum duration of some National Security Determinations and aligning terrorism-related retention rules. Lawful biometric use therefore depends on both the general PoFA framework and any specific terrorism or national security provisions that apply [50].

International biometric exchanges must meet both legal and quality requirements. The Accreditation of Forensic Service Providers Regulations 2018, as amended in 2019, require appropriate accreditation for

[1] https://assets.publishing.service.gov.uk/media/65e1ab9d2f2b3b00117cd803/Attorney_General_s_Guidelines_on_Disclosure_-_2024.pdf.

forensic laboratory activity linked to the Prüm exchange system. This reinforces that DNA and fingerprint exchange should be supported by assessed and quality-assured processes.

The EU–UK Trade and Cooperation Agreement 2020 provides the current framework for continuing biometric exchange with EU Member States after Brexit. It allows DNA, fingerprint, and vehicle registration data to be checked through a Prüm-style 'hit/no hit' process. This means a technical match can be identified first, while any later request for personal or case information must follow separate legal steps. The process can support identification and crime-linkage work, but only where data quality, lawful processing, and forensic reliability are maintained.

Data protection and disclosure are also essential. Under the [18], law-enforcement processing must be lawful, fair, necessary, proportionate and secure. For biometric material, this requires clear justification, access controls, retention decisions, and safeguards. Disclosure duties also require investigators to retain and reveal material that may be relevant to the case, including unused marks, laboratory notes, contamination records, image-processing material, comparison records, or alternative interpretations. This enables the defence to test the evidence properly and supports a fair trial [4, 15, 18].

## 12.3 Effects of Regulation on DNA, Fingerprints and Facial Image Processing

For DNA, regulation strengthens contamination control, validated workflows, and careful interpretation of source-and activity-level findings. This is especially important for partial, mixed, or low-level profiles where transfer, background DNA, and uncertainty must be addressed clearly [25, 26].

For fingerprints, regulation supports structured quality systems for examination, verification and documentation. It also strengthens competence standards, validated enhancement methods, and clearer reporting of limitations, particularly where latent marks are incomplete or poor quality [22, 47].

For facial images, specific legislation remains more limited, but governance is increasingly important as retrospective, live, and operator-initiated facial recognition develop. Reliability depends on image quality, database selection, threshold settings, algorithm performance, and human review. A facial recognition match should therefore be treated as an interpreted output, not a self-explanatory conclusion [32, 33].

Recent consultation activity indicates that facial identification may move towards broader regulated use. Two key areas under consideration are a retention regime for policing images, potentially aligned with existing DNA and fingerprint frameworks, and stronger oversight of facial recognition technology in policing. The final shape of any future framework remains uncertain. For forensic practice, this strengthens the need for safeguards around proportionality, bias, transparency, validation, explainability and public trust, particularly where outputs may influence investigations or court proceedings [39, 13, 14, 59].

### *12.3.1 Current Representation of AI in the Forensic Legal and Governance Landscape*

AI offers a significant opportunity to improve forensic science across different biometric material types, including DNA, fingerprints, and facial images. However, current forensic legislation and governance do not directly address the use of AI within forensic processes. In the author's view, AI should be introduced slowly and cautiously because of its potential impact on the criminal justice system. With careful design, development, and implementation, AI could make biometric analysis faster, more consistent, and more efficient. However, standards must remain high, and specialist forensic expertise must continue to guide its use so that appropriate safeguards are in place and the risk of miscarriages of justice is minimised.

## 12.4 Potential AI Use in DNA, Fingerprints and Facial Images

In DNA analysis, AI may help with tasks such as assessing profile quality, identifying alleles, interpreting mixtures, supporting kinship analysis, and examining complex or degraded samples (see Box 1 for types of DNA profiles).

*Box 1: Explanation of Types of DNA Profiles*

DNA profiling is a scientific method used to examine specific areas of a person's DNA, most commonly short tandem repeats (STRs), to produce a profile that can be compared with DNA from a person, item or crime scene. A person usually has two DNA components represented as peaks at each area of the DNA analysed, one is inherited from the mother and second from the father.

- **A good single-source profile** usually comes mainly from one person and shows clear, strong and balanced peaks at each area, making it more straightforward to interpret.
- **A low-level profile** contains only a small amount of DNA, so some peaks may be missing, weak or uneven, which increases the risk of uncertainty.
- **A degraded profile** occurs when DNA has broken down because of time, heat, moisture, sunlight, chemicals, or microbial activity. This can make longer DNA fragments harder to detect and may result in partial/incomplete DNA profiles.
- **A mixed profile** contains DNA from two or more people, so the scientist must consider which alleles may belong to each contributor. A major–minor mixture is a type of mixed profile where one person's DNA is much stronger than another's; this may help identify the main contributor, but the weaker contributor can be difficult to interpret if their peaks are small, overlap with the major contributor, or fall close to the reporting threshold.

These examples show that DNA interpretation is not simply reading peaks from a graph: it requires validated methods, awareness of profile quality, and careful scientific judgement [6, 7, 30, 31, 49, 55, 58, 57].

Studies suggest that these tools may improve speed and consistency, especially where profiles are mixed, low-level, or difficult to interpret. The main risks are that automated outputs may be misunderstood, treated as more certain than they are, or used without sufficient explanation, validation, or awareness of population and dataset bias. There is also a risk of false positive results or hallucinations with use of AI which may affect the results or interpretation. There may also be challenges in explaining probabilistic or AI-assisted findings clearly in court [9, 51, 53, 63].

In fingerprint workflows, AI may support image enhancement, ridge and minutiae detection, candidate-list generation, and crime scene mark searching. AI applications for these areas may be valuable where marks are partial, smudged, or difficult to search manually due lack of friction ridge detail. However, utilising AI includes risks for consideration in terms of poor performance on low-quality marks, limited transparency, dataset bias, and weak transfer from research settings to real operational conditions. Candidate lists or rankings (see Box 2 for explanation on searching and rank lists) should therefore be treated as investigative support, not as proof of identity, with expert examiners remaining responsible for interpretation and reporting [34, 35].

*Box 2: Finger Mark Search Explanation*

Crime scene marks, often called latent fingerprints, Friction Ridge Detail or finger-marks, are usually partial, faint, or distorted impressions left on surfaces. Once recovered and photographed or scanned, the mark can be prepared for searching by improving the image quality and identifying ridge detail and minutiae, such as ridge endings and bifurcations. An Automated Fingerprint Identification System (AFIS) then compares the mark against stored reference fingerprints and produces a ranked candidate list of possible matches, usually based on similarity scores and ranked in this order. This list does not prove identity; it is only an investigative search result. A trained fingerprint expert must manually examine the crime scene mark against the suggested reference prints, assess the quality and quantity of ridge detail, consider any differences, and reach a conclusion using a structured comparison and verification process. This manual expert verification is essential because poor-quality marks, distortion, background noise, limited ridge detail, and database ranking effects can influence automated search results and may create false or misleading candidate suggestions [11, 20, 36, 61].

For facial images, AI may support retrospective searches, live facial recognition, operator-initiated checks, image triage, and candidate-list generation. The research identifies not only clear investigative benefits, but also important risks. These include false positives and false negatives, reduced performance with poor-quality images, demographic performance differences, weak safeguards, surveillance expansion, and reduced public trust. Image quality issues such as blur, low resolution, pose, and lighting can affect performance, and human review may not be enough if users place too much confidence in algorithmic results [2, 28, 46].

Across all three biometric areas, the academic literature points to the same governance lesson: AI should be used to support, not replace, expert decision-making. Before use, systems should be independently validated, locally verified, monitored for bias and error, and supported by clear records explaining the data, model, thresholds, limitations and human review applied. This is particularly important where outputs may affect suspect identification, charging decisions, evidential interpretation or disclosure [28, 29, 35, 53]. To summarise, AI could support DNA, fingerprint and facial-image analysis by improving speed and consistency, but it must be validated, transparent and used to support, not replace expert forensic judgement.

## 12.5 Governance of AI in Forensics

There is no single UK statute that governs AI in policing or in forensics. Instead, police use of AI must fit within public law, data protection law, human rights law, sector guidance, and operational oversight. This means forces must be able to show that any AI deployment is lawful, necessary, proportionate, and fair, particularly where it involves personal or biometric data, or it may affect rights such as privacy, expression, assembly and non-discrimination [18, 41, 52].

Additional controls may apply where AI is used for surveillance, communications data, or biometric identification. The Home Office

Biometrics Strategy emphasises privacy, ethics, impact assessment, oversight, and standards, while wider UK AI policy promotes safety, transparency, fairness, accountability, and contestability through a principles-based approach [19, 38, 44]; (Department for Science, Innovation and Technology 2023).

Practical governance tools for public bodies include the Algorithmic Transparency Recording Standard and the AI Playbook.[2] These support transparency, human control, risk management, lifecycle assurance, and legal compliance. The ICO also expects organisations using AI to evidence accountability through governance records, data protection impact assessments, clear responsibilities, and ongoing oversight [32, 33, 42, 43].

Oversight is shared across several bodies and processes. The Biometrics and Surveillance Camera Commissioner has supported lawful and accountable use of biometric and facial recognition technologies, while police forces remain responsible for deployment decisions, records, impact assessments, and internal governance. Data ethics committees may also support review of data-driven policing projects, although there is not yet a single national approval system [8, 52].

Parliamentary scrutiny has also raised concerns about legal clarity, particularly for live facial recognition. The House of Lords Justice and Home Affairs Committee questioned whether deployments had a sufficiently clear legal basis, while government statements recognise both the need for clearer rules and the potential public protection benefits of AI-enabled policing [40, 39].

## 12.6 Proposing an Operational Framework for AI in Biometric Systems

AI used in biometric forensic work should be treated as a high-impact capability because it may influence investigations, evidential interpretation, and criminal justice outcomes. The author considers the proposed framework is distinct from more generic policing applications of AI because it seeks to integrate established governance within a mature scientific discipline with the flexibility required to respond to rapidly evolving

[2] https://assets.publishing.service.gov.uk/media/67aca2f7e400ae62338324bd/AI_Playbook_for_the_UK_Government__12_02_.pdf.

technologies. Governance should therefore combine forensic requirements on validation, competence, quality management, and reporting with public-sector AI duties on legality, transparency, accountability, human oversight, and contestability. This is especially relevant to AI tools that support DNA interpretation, fingerprint comparison, facial-image searching, or biometric triage [24, 32, 43, 39].

Implementation should begin with a clear lawful purpose, documented necessity and proportionality, and evidence that the system improves capability or efficiency without undermining scientific validity, fairness, explainability, or individual rights. AI outputs should not be treated as evidence in themselves. They should be assessed according to their function, evidential significance, and the level of human review applied [1, 18, 29, 41]; (Bridges v South Wales Police [2020] EWCA Civ 1058[3]).

### 12.6.1 *Assurance, Human Decision-Making, and Disclosure*

Assurance should bring together scientific validation, quality management, legal governance, and operational oversight. Organisations should keep records of intended use, validation, verification, performance, datasets, thresholds, training, audit logs, version changes, security checks, incidents, and system limitations. Extra caution is needed for opaque or 'black-box' models, particularly where outputs may affect evidential interpretation or charging decisions [1, 29, 32, 39]. Given the rapid and evolving nature of AI technologies, particular challenges may arise in demonstrating compliance with established assessment regimes such as ISO/IEC 17025 and in applying the Code consistently, especially in relation to repeatability and reproducibility. These issues require careful consideration, and regulatory frameworks may need to continue developing to accommodate the distinctive characteristics of AI-enabled forensic tools.

AI should support, not replace, expert forensic judgement. Procedures should define who may use the system, what outputs can be relied on, when expert escalation is required, and how uncertainty is presented. Rankings, scores, or similarity measures should not be treated as conclusive without context. Training should also cover system limits,

[3] *Bridges v South Wales Police* [2020] EWCA Civ 1058. Microsoft Word - R (Bridges) -v- CC South Wales _ors Judgment.docx.

cognitive bias, disclosure duties, and the difference between intelligence and evidential use [29, 32, 39].

Where AI influences an investigative decision, forensic result, or evidential report, its role should be documented and disclosed where they are deemed significant. Reports should distinguish between system-generated outputs, expert observations, and evaluative conclusions, and should avoid overstating objectivity. This supports admissibility, weight, and fair challenge in criminal proceedings [33, 39] (Bridges v South Wales Police [2020] EWCA Civ 1058).

### *12.6.2 Proposed Phased Roadmap for Implementation*

**Phase 1: Governance readiness**: Establish a senior accountable owner, multidisciplinary governance group, decision log, legal and ethical assessment template, and links to forensic quality management and data governance processes.

**Phase 2: Use-case selection**: Prioritise narrow, high-value use cases where the task is well defined and human oversight is straightforward, such as triage support rather than fully automated evidential conclusions.

**Phase 3: Data and validation preparation**: Curate datasets, define metrics, agree thresholds, and complete independent validation and local verification.

**Phase 4: Pilot deployment**: Run a time-limited pilot with trained users, enhanced audit, CPS engagement where relevant, and pre-agreed evaluation criteria.

**Phase 5: Controlled operationalisation**: Move into business-as-usual only if scientific, legal and operational criteria are met. Monitoring, disclosure readiness and governance review should continue throughout use.

**Phase 6: Review and retirement**: Periodically review continued necessity, performance, fairness, and legal fit. Systems that can no longer be justified, supported, or validated should be retired.

## 12.7 Conclusion

AI could improve biometric forensic science in England and Wales by helping practitioners manage complex data more quickly and consistently across DNA, fingerprints, and facial images. However, it should not be seen simply as a technical improvement. Where AI supports searching, comparison, interpretation, or reporting, it should be governed as a high-impact forensic capability because its outputs may affect investigations, evidential assessments, and criminal justice outcomes. This means combining forensic requirements on validation, competence, quality management, reporting, and disclosure with AI governance principles such as legality, transparency, fairness, accountability, contestability, and meaningful human oversight [32, 33, 43, 39].

The governance landscape is fragmented but connected. DNA and fingerprints are more clearly embedded within forensic regulation and statutory retention rules, while facial identification remains less consistently regulated at the recovery and initial processing stage. Even so, all biometric techniques raise common issues around reliability, proportionality, privacy, equality, bias, database quality, disclosure, and public trust. The decision in *Bridges v South Wales Police* shows that biometric technologies, particularly facial recognition, require a clear legal basis, adequate safeguards and proper consideration of human rights, data protection, and equality duties [18, 37, 41, 39] (Bridges v South Wales Police [2020] EWCA Civ 1058).

The proposed way forward is incremental, auditable, and evidence led. AI systems should only be introduced where the use case is clear, the lawful purpose is documented, and the system has been independently validated and locally verified for the intended forensic context. Particular caution is needed for opaque or 'black-box' models because criminal justice users, experts, lawyers, and courts must be able to understand the basis, limits, and significance of any AI-assisted output. AI should therefore support, rather than replace, expert judgement, and reports should separate system-generated outputs from expert interpretation and evidential conclusions [1, 4, 15, 29].

Overall, the strongest model is a combined governance approach rather than a separate AI regime outside forensic science. This approach should bring together forensic quality standards, data protection safeguards, human rights principles, operational accountability, and lifecycle assurance. The phased roadmap provides a practical route for doing this through governance readiness, careful use-case selection, data and validation preparation, controlled piloting, monitored operational use, and periodic review or retirement. If applied consistently, AI could strengthen biometric forensic capability while protecting scientific integrity, individual rights, public confidence, and the fairness of the criminal justice system [29, 32, 43, 39].

## References


1. O. Ali et al., The effects of artificial intelligence applications in educational settings: Challenges and strategies. Technol. Forecast. Soc. Chang. **199**(123076), 123076–123076 (2024). https://doi.org/10.1016/j.techfore.2023.123076
2. V. Amador-Lankster, Facial recognition in policing: how algorithmic bias targets people of color. Undergrad. Law Rev. UC San Diego **3**(1) (2025). https://doi.org/10.5070/lr3.47403. Accessed 16 July 2026
3. D.R. Ashbaugh, *Quantitative-Qualitative Friction Ridge Analysis* (CRC Press, 1999)
4. Attorney General's Office, *Attorney General's Guidelines on Disclosure* (Attorney General's Office. Attorney_General_s_Guidelines_on_Disclosure_-_2024.pdf, London, 2024)
5. N. Bacci et al., Forensic facial comparison: current status, limitations, and future directions. Biology **10**(12), 1269 (2021). https://doi.org/10.3390/biology10121269
6. D.J. Balding, Evaluation of mixed-source, low-template DNA profiles in forensic science, in *Proceedings of the National Academy of Sciences*, vol. 110, no. 30, pp. 12241–12246 (2013). https://www.pnas.org/content/pnas/110/30/12241.full.pdf. https://doi.org/10.1073/pnas.1219739110. Accessed 16 July 2026
7. F.R. Bieber et al., Evaluation of forensic DNA mixture evidence: protocol for evaluation, interpretation, and statistical calculations using the combined probability of inclusion. BMC Genet. **17**(1) (2016). https://doi.org/10.1186/s12863-016-0429-7
8. Biometrics and Surveillance Camera Commissioner, *Annual Report 2023–2024* (Home Office, London, 2024) Biometrics and Surveillance Camera Commissioner's annual report 2023 to 2024 (accessible) - GOV.UK
9. D. Brancato, E. Coniglio, F. Bruno, S. Treccarichi, M. Vinci, F. Calì, S. Saccone, C. Federico, Beyond STRs: Integrative forensic genomics from MPS to genetic genealogy and AI-based prediction. Genes **17**(5), 580 (2026)
10. J.M. Butler, *Advanced Topics in Forensic DNA Typing: Methodology* (Academic Press, 2011)
11. K. Cao, A.K. Jain, Automated latent fingerprint recognition. IEEE Trans. Pattern Anal. Mach. Intell. **41**(4), 788–800 (2019). https://doi.org/10.1109/tpami.2018.2818162
12. C. Champod, C.J. Lennard, P. Margot, M. Stoilovic, *Fingerprints and other ridge skin impressions*. Google books (2016). Fingerprints and other ridge skin impressions - Christophe Champod, Chris J. Lennard, Pierre Margot, Milutin Stoilovic - Google books

13. College of policing, Live facial recognition technology guidance published, in *College of Policing* (2022). https://www.college.police.uk/article/live-facial-recognition-technology-guidance-published. Accessed 16 July 2026
14. College of Policing, Live Facial Recognition Technology Guidance Published. [online] College of Policing (2022). Available at: https://www.college.police.uk/article/live-facial-recognition-technology-guidance-published
15. Criminal Procedure and Investigations Act 1996. London: The Stationery Office. Criminal Procedure and Investigations Act 1996
16. Crown Prosecution Service, *Forensic Science Regulator Code of Practice: Legal Guidance* (Crown Prosecution Service, London, 2023) Available online: https://www.legislation.gov.uk/ukpga/2021/14/contents
17. DSTL, *Fingermark Visualisation Source Book V 3.0 The Scientific Rationale Behind the Processes Within Fingermark Visualisation Manual Second Edition 2022* (2022)
18. Data Protection Act 2018. London: The Stationery Office. Data Protection Act 2018
19. Department for Science, Innovation & Technology, *A pro-innovation Approach to AI regulation: Government Response*. [online] GOV.UK (2024). Available at: https://www.gov.uk/government/consultations/ai-regulation-a-pro-innovation-approach-policy-proposals/outcome/a-pro-innovation-approach-to-ai-regulation-government-response
20. U.U. Deshpande et al., End-to-end automated latent fingerprint identification with improved DCNN-FFT enhancement. Front. Rob. AI 7 (2020). https://doi.org/10.3389/frobt.2020.594412
21. Facial Identification Scientific Working Group, *Facial Comparison Overview and Methodology Guidelines*. Version 1.0 (2019). FISWG - Facial Comparison Overview and Methodology Guidelines - Version 1.0 - 2019.10.25
22. Forensic Science Regulator, *Codes of Practice and Conduct Fingerprint Comparison Appendix: FSR-C-128* (2014). Available online: https://www.gov.uk/government/publications/forensic-science-activities-statutory-code-of-practice-version-2/forensic-science-activities-statutory-code-of-practice-version-2-accessible
23. Forensic Science Regulator (2015). *Annual Report November 2015*. Available online: https://www.gov.uk/government/publications/forensic-science-regulator-annual-report-2015
24. Forensic Science Regulator (2021). *Forensic Science Regulator Act 2021*. Available online: https://www.legislation.gov.uk/ukpga/2021/14/contents

25. Forensic Science Regulator, *DNA Contamination Controls: Laboratory. FSR-GUI-0018* (2023). https://www.gov.uk/government/publications/dna-contaminationcontrols-laborator
26. Forensic Science Regulator, *Guidance. Contamination Controls: Scene of Crime, 415 FSR-GUI-0016* (2023). https://www.gov.uk/government/publications/crime-scene-dna-anti-contamination-guidance/contamination-controls-scene-of-crime-accessible
27. Forensic Science Regulator, *Guidance: Forensic science activities: statutory code of practice - version 2 (accessible)* (2025). Available online: https://www.gov.uk/government/publications/forensic-science-activities-statutory-code-of-practice-version-2/forensic-science-activities-statutory-code-of-practice-version-2-accessible
28. P. Fussey et al, *Facial Recognition Surveillance* (2025)
29. S.A. Garrett, C. Rudin, Shibboleth authentication request. Oclc.Org (2026) https://access-heinonline-com.hallam.idm.oclc.org/hol/page?collection=journals&handle=hein.journals/clqv109&id=574&men_tab=srchresults
30. P. Gill et al., Forensic application of DNA 'fingerprints. Nature **318**(6046), 577–579 (1985)
31. P. Gill et al., An investigation of the rigor of interpretation rules for STRs derived from less than 100 Pg of DNA. Forensic Sci. Int. **112**(1), 17–40 (2000). https://doi.org/10.1016/s0379-0738(00)00158-4. Accessed 16 July 2026
32. Government Digital Service, GDS 2025 Artificial intelligence playbook for the UK government. (2025)
33. Government Digital Services, GDS 2025. GOV.UK (2023). https://www.gov.uk/government/collections/algorithmic-transparency-recording-standard-hub
34. S.A. Grosz, A.K. Jain, *Latent Fingerprint Recognition: Fusion of Local and Global Embeddings* (2023). [online] arXiv.org. Available at: https://arxiv.org/abs/2304.13800
35. C. Herke, Automated fingerprint identification: The role of artificial intelligence in crime scene investigation. Forensic Sci. **6**(1), 6–6 (2026). https://doi.org/10.3390/forensicsci6010006
36. R.A. Hicklin et al., Accuracy and reproducibility of latent print decisions on comparisons from searches of an automated fingerprint identification system. Forensic Sci. Int., 112457 (2025). https://doi.org/10.1016/j.forsciint.2025.112457
37. Home Office, *Protection of Freedoms Act 2012: DNA and Fingerprint Provisions* (London: Home Office, 2013). Available at: https://www.legislation.gov.uk/ukpga/2012/9/contents

38. Home Office, *Biometrics Strategy* (Home Office, London, 2018) Available at: https://www.gov.uk/government/publications/home-office-biometrics-strategy
39. Home Office, Consultation on a new legal framework for law enforcement use of biometrics, facial recognition and similar technologies (accessible). GOV.UK (2025). https://www.gov.uk/government/consultations/legal-framework-for-using-facial-recognition-in-law-enforcement/consultation-on-a-new-legal-framework-for-law-enforcement-use-of-biometrics-facial-recognition-and-similar-technologies-accessible
40. House of Lords Justice and Home Affairs Committee, *Technology Rules? The Advent of New Technologies in the Justice System* (House of Lords, London, 2024) Technology Rules? The advent of new technologies in the justice system
41. Human Rights Act (1998). *Human Rights Act 1998*. [online] Legislation. gov.uk. Available at: https://www.legislation.gov.uk/ukpga/1998/42/contents
42. Information Commissioner's Office, *Guidance on AI and Data Protection* (2023). Available at: https://ico.org.uk/for-organisations/uk-gdpr-guidance-and-resources/artificial-intelligence/guidance-on-ai-and-data-protection/
43. Information Commissioner's Office, *AI and Data Protection: Accountability and Governance Guidance* (Information Commissioner's Office, Wilmslow, 2025) Available at: https://ico.org.uk/for-organisations/uk-gdpr-guidance-and-resources/artificial-intelligence/guidance-on-ai-and-data-protection/what-are-the-accountability-and-governance-implications-of-ai/
44. Investigatory Powers Act, *London: The Stationery Office* (Legislation.gov.uk, Available, 2016) at: https://www.legislation.gov.uk/ukpga/2016/25/contents
45. D.-O. Jaquet-Chiffelle, E. Casey, A formalized model of the trace. Forensic Sci. Int. **327**, 110941 (2021). https://doi.org/10.1016/j.forsciint.2021.110941
46. N. Lynch, Facial recognition technology in policing and security—case studies in regulation. Laws **13**(3), 35 (2024). https://www.mdpi.com/2075-471X/13/3/35. https://doi.org/10.3390/laws13030035. Accessed 16 July 2026
47. N. Martins et al.: Fingerprint recognition in forensic scenarios. Sensors (Basel, Switzerland) **24**(2), 664 (2024). https://www.ncbi.nlm.nih.gov/pmc/articles/PMC10819264///. https://doi.org/10.3390/s24020664. Accessed 16 July 2026

48. E. Mistek et al., Toward Locard's exchange principle: recent developments in forensic trace evidence analysis. Analytical Chemistry **91**(1), 637–654 2018. https://pubs.acs.org/doi/full/10.1021/acs.analchem.8b04704. https://doi.org/10.1021/acs.analchem.8b04704. Accessed 16 July 2026
49. S. Nakao et al., Effective use of the degradation index from human DNA quantification kits to improve STR and Y-STR profiling. Forensic Sci. Int. Genet. 103342–103342 (2025). https://doi.org/10.1016/j.fsigen.2025.103342
50. Office of the Biometrics and Surveillance Camera Commissioner (2020). *Annual Report 2020*. Available online: https://assets.publishing.service.gov.uk/media/61a10e698fa8f50381640393/E02669527_Biometrics_Commissioner_ARA_2020_Text_Accessible.pdf
51. D. Primorac et al., When algorithms testify: Artificial intelligence-driven DNA analysis, evidentiary standards, and criminal justice reform. Croat. Med. J. **67**(3), 203–218 2026. https://pmc.ncbi.nlm.nih.gov/articles/pmc13247746/, https://doi.org/10.3325/cmj.2026.67.203. Accessed 16 July 2026
52. G. Sartor, Artificial intelligence and human rights: Between law and ethics. Maastricht J. Eur. Comp. Law **27**(6), 705–719 (2020). https://doi.org/10.1177/1023263x20981566. Accessed 16 July 2026
53. F. Sessa et al., Artificial intelligence and forensic genetics: current applications and future perspectives. Appl. Sci. **14**(5), 2113 (2024). https://www.proquest.com/docview/2955455349?accountid=13827&parentsessionid=9khqml%2byy7z%2fe5awfav35bcq88rzcr2ka%2btakmuux74%3d&pq-origsite=primo&sourcetype=scholarly%20journals. https://doi.org/10.3390/app14052113
54. S.V. Stevenage, C. Pitfield, Fact or friction: examination of the transparency, reliability and sufficiency of the ACE-V method of fingerprint analysis. Forensic Sci. Int. **267**, 145–156 (2016). 10.1016/j.forsciint.2016.08.026. Accessed 16 July 2026
55. D. Taylor et al., The interpretation of single source and mixed DNA profiles. Forensic Sci. Int. Genet. 7(5), 516–528 (2013). https://doi.org/10.1016/j.fsigen.2013.05.011
56. T. Trejos et al., Scientific foundations and current state of trace evidence—a review. Forensic Chem. **18**, 100223 (2020). https://doi.org/10.1016/j.forc.2020.100223
57. T. Tvedebrink et al., Statistical model for degraded DNA samples and adjusted probabilities for allelic drop-out. Forensic Sci. Int. Genet. **6**(1), 97–101 (2012). https://doi.org/10.1016/j.fsigen.2011.03.001
58. T. Tvedebrink et al., Estimating drop-out probabilities of STR alleles accounting for stutters, detection threshold truncation and degradation.

Forensic Sci. Int.: Genet. Suppl. Ser. **4**(1), e51–e52 (2013). https://www.sciencedirect.com/science/article/abs/pii/s1875176813000279. https://doi.org/10.1016/j.fsigss.2013.10.026. Accessed 16 July 2026
59. UK Parliament, Police: facial recognition technology - Hansard - UK Parliament. *Parliament.Uk* (2026). https://hansard.parliament.uk/lords/2025-07-02/debates/9b397063-6bbc-4091-b349-e40df717a654/policefacialrecognitiontechnology. Accessed 16 July 2026
60. N.S. Udogadi et al., Forensic DNA profiling: autosomal short tandem repeat as a prominent marker in crime investigation. Malays. J. Med. Sci. **27**(4), 22–35 (2020). https://pmc.ncbi.nlm.nih.gov/articles/PMC7444828/. https://doi.org/10.21315/mjms2020.27.4.3. Accessed 16 July 2026
61. B.T. Ulery et al., Accuracy and reliability of forensic latent fingerprint decisions, in *Proceedings of the National Academy of Sciences*, vol. 108, no. 19, 25, pp. 7733–7738 2011. https://www.ncbi.nlm.nih.gov/pmc/articles/PMC3093498/. https://doi.org/10.1073/pnas.1018707108. Accessed 16 July 2026
62. D. White et al.: Passport Officers' Errors in Face Matching. PLoS ONE **9**(8), e103510 (2014). https://doi.org/10.1371/journal.pone.0103510
63. R. Zhang, Artificial intelligence in forensic genetics: applications and ethical challenges. J. Transl. Genet. Genomics **9**(4), 359–367 (2025). https://www.oaepublish.com/articles/jtgg.2025.76. https://doi.org/10.20517/jtgg.2025.76. Accessed 16 July 2026

**Shazia Khan** has more than 25 years of experience in law enforcement and biometrics and has established a distinguished career at the forefront of innovation, governance, and public service. Throughout her career, she has held a range of senior leadership positions, driving strategic change and shaping policy to ensure the ethical, proportionate, and responsible use of emerging technologies. Her expertise has been instrumental in advancing best practices and fostering public trust in the deployment of biometric solutions. Recognised for her thought leadership in the field, she actively contributes to numerous national and international biometrics working groups, helping to influence policy, standards, and the future direction of biometric technologies worldwide.

CHAPTER 13

# AI and Serious and Organised Crime

*Christopher Rowley, Luke Bates, and Gabriela Boneva*

## 13.1 Introduction

Serious and organised crime groups (SOCGs) have played a significant role in history, from the Cilician Pirates who operated in the Roman Republic who allegedly kidnapped Julius Ceasur and held him captive for ransom, to Al Capone in America. SOCG's have demonstrated dexterity, flexibility, and a willingness to try new ideas, showcasing their inherent adaptability. They do not work within any framework, and there are no moral or values which guide them. Their offending is often for personal financial gain or self-gratification, with no consideration for the victims or individuals that are affected by their criminal exploits. Their chameleon-like tendencies have allowed them to quickly adopt and adapt to new technology, looking to see how it can assist in maximising their effectiveness both locally and—in many cases—internationally, whilst bolstering their profits.

Through necessity they are constantly reviewing their working practices to maintain and maximise their position and profits, and to stay one step ahead of the police to avoid arrest and the disruption of their illegal

C. Rowley (✉) · L. Bates · G. Boneva
CENTRIC, Sheffield Hallam University, Sheffield, UK
e-mail: christopher.rowley@shu.ac.uk



B. Akhgar and P. S. Bayerl (eds.), *Accountability for Artificial Intelligence in Policing and Security*, Security Informatics and Law Enforcement, https://doi.org/10.1007/978-3-032-41284-3_13

activities. A recent example of this was SOCGs' use of the Encrochat—a secure messaging system installed on modified mobile phones—which criminals believed could not be intercepted. The SOCGs' using Encrochat were identified and dismantled [8, 9]; the subsequent operations resulted in 6558 arrests, €739.7 million being seized, along with firearms, houses, boats, and significant quantities of drugs.

Encrochat is but one example of SOCGs using technology, and their move into the use of AI is well documented. The Police must consider how they can legitimately use AI and the intelligence they gain from their investigations to predict how SOCG's are using technology or will use technology in their future criminal endeavours. The system may identify issues that have been missed by busy operational officers and staff, or it may suggest questions that should be considered when considering new technologies. The adoption of new AI systems will assist policing and law to remain in step with SOCG and, if the organisations invest time and allow their officers/staff to be innovative, possibly gain ground on the SOCG.

Police adopting and using the correct AI system in the investigation of SOCG criminality will become mainstream in the coming years through necessity, because of the complexity of large-scale forensic digital data, cross border criminality and jurisdictions, and the criminal adoption and adaption of AI. Harm caused to individuals and the potential of imminent threats to their lives also contribute to the highlighted problems, as their prevention often requires a number of specialised resources, which come at a significant cost to the organisation. The problems faced by policing and the increasing costs of these developments demand new and improved ways of working, which are expeditious, flexible, and effective, whilst demonstrating the desired benefits such as increased efficiency and reduced workload demand on individual investigators through automation, and the ability to identify criminal patterns indiscernible to the human eye.

This chapter will provide an insight into the SOCG world of criminality, which is sophisticated, complex, flexible, innovative, and ruthless with the added problem of geographically borderless and criminal exploitation of AI. The combination of each of these elements can cause significant issues to finite investigative resources; that said, the correct AI system can, and will, significantly reduce the demand on the investigator. It will speed up the investigation whilst identifying links which otherwise may have been missed, and as previously detailed, it will possibly identify new effective and efficient ways of conducting investigations.

## 13.2 Snapshot: The Organised Crime Threat

Serious and organised crime represents one of the greatest threats to our society, both in terms of values, corruption, and violence, as well as the impact that the operation of vast underground illegal and illicit markets has on our economy with the funnelling of money away from legitimate enterprise and commerce. There is no greater enabler of serious and organised crime than the drugs market and the trafficking of illegal and illicit substances across great distances in pursuit of profit. Through the cultivation, manufacturing, trafficking, and sale of drugs and their precursors, criminals also engage with a plethora of related crimes, such as money laundering, corruption, and serious violence. To make matters more complex, the advent of new, digital forms of anonymised communication, such as the dark web and encrypted platforms, as well as exploitable legal loopholes, means that traditional SOCGs are also under competition by lone actors operating within this space, whose Modus Operandi (MO) and profiles may be unknown to law enforcement.

There is therefore a demand for innovations in intelligence-based and investigatory capabilities that can harness the latest developments in AI technology to fight this illusive threat. The ability to track the drugs trafficking chain, both online and offline, is crucial for law enforcement to be able to disrupt and take down the vast criminal networks and agents who continue to flood the market with their product.

Drugs trafficking is a cross-border crime, with links between criminal networks spanning continents and criminals exploiting both legitimate and illegal transportation channels, meaning that trans-national interagency cooperation and intelligence sharing is a must in order to efficiently thwart efforts to take advantage of differing jurisdictions and legal environments. The increasing shift from physical to online marketplaces by drug criminals further reduces the geographical restraints of traditional face-to-face drug dealing, providing drug vendors with expanded market reach and a highly reduced risk of detection by law enforcement. This ever-growing illicit use of online spaces for drug trafficking puts innocent individuals working in postal and courier services at risk of criminal exploitation, who, often unaware of it, end up distributing substances within and across countries. This criminal exploitation of legitimate commercial transit infrastructures via postal and courier services is often characterised with polydrug trafficking, where relatively small quantities of multiple substances are distributed together at retail level [10].

Research from Scotland sheds light on the geographic patterns of intercepted postal drug deliveries sold online [14], which reported that the majority of the drug packages were destined for highly populated Scottish urban centres, such as Glasgow, Dundee, Edinburgh, and Aberdeen, likely due to the greater Internet connectivity and higher levels of access to services associated with bigger urban areas. However, the same research also reported observing higher than expected levels of drug parcels destined to more remote areas, albeit a following spatial clustering analysis showed that the drug parcels sent out to the most remote and rural areas of the country were still associated with good access to services, with the packages therefore usually destined to addresses within larger settlements of the rural areas. These results suggest that, from a drug-consumer standpoint, Internet connectivity and access to postal and courier services are the leading enablers of purchasing drugs online and getting them delivered through legitimate channels.

Besides the connection observed between online drugs trafficking and the exploitation of legitimate transportation channels, increasing globalisation, defined by better connected communication and transportation networks, also links to expanded opportunities not only beneficial to the public, but criminally exploited for drugs trafficking. Seaports and maritime shipping ports continue to be exploited for bulk trafficking of illegal drugs, often involving infiltration, corruption, and intimidation to personnel working in these sectors [7]. Besides maritime routes, drug traffickers also employ continuingly evolving techniques to evade detection, including combining land and sea routes, as well as using EU and non-EU transhipment points to conceal illegal trafficking and challenge the detection of their criminal activities.

Globalisation also fuels the development of interconnected communication networks, which in their own right contribute to innovative new drugs trafficking approaches and fuel the risk of a global overdose crisis. Cyberspace breaks down geographical limits, enabling individuals and networks to connect and trade illicit goods from any point of the globe–a concept in total disagreement with traditional SOCGs, which characterised themselves in high territoriality, but highlighting criminal adaptability and exploitation skills [13]. Extending from surface websites and social media platforms to encrypted messaging platforms and dark web markets, illicit drug dealing has spread across the online landscape, calling for sophisticated technologies that can support police efforts in tackling organised crime whilst handling the vastness of heterogeneous data at hand. Furthermore, cryptocurrency is also now becoming a well-known method for payment of drugs online, which adds to the layers of anonymity provided, with sophisticated criminals reportedly making use of cross-chain transactions, mixers, tumblers, and deploy stable coins to obfuscate their online illicit activities [1].

A key trait of serious and organised crime is the likely involvement of individuals and networks in multiple criminal activities to either support their existing criminal ventures or expand their profits. Drugs trafficking, arguably being the most violent sector of the illegal economy [4], is often associated with other forms of serious crime, such as money laundering, corruption, and illicit goods cross-trafficking. The violence associated with drug trafficking implies the need for actors to arm themselves for self-protection and intimidation, increasing the spread and number of illegal firearms and associated violence, with between 28% and 44% of all firearm seizures in Europe occurring in the context of drugs trafficking [6]. The nexus between illegal firearms and drugs trafficking is further reinforced through cross-trafficking, with SOCGs using the same routes to traffic both drugs and firearms or using illicit goods as a form of cashless payment between different groups.

Some SOCGs involved with drugs trafficking transfer their criminal activities to environmental crimes, predominantly due to the lower risk posed by illegal wildlife trades than illegal drugs trades, linked to less strict environmental and wildlife crime laws. Such crime divergence is not a new phenomenon, with similar events being observed in the 1970s and 1980s when criminals shifted to committing drug crimes from other more serious types of crime due to the legality around drugs being less strict back then [16]. Due to the greater risk-to-reward ratio, drug traffickers also exploit the less strict sanctions for wildlife trafficking to conceal illicit drugs in parallel trafficking activities. Combined trafficking–a type of parallel trafficking–of wildlife and drugs, for instance reportedly not only increases criminal profits but also decreases the risk of drug checks and detection [19]. Another concealment method for drugs smuggling is the so-called camouflage trafficking, where drugs are hidden in legal commodities—for instance, liquid cocaine lining the plastic bags of live fish or the surgical implantation of legal exotic animals with cocaine [18].

The globality of drugs trafficking and the adaptability of SOCGs calls for continuous development, adaptation, and utilisation of modern technologies for law enforcement, equipping authorities with the tools required for the arms race against illegal and illicit drugs and serious and organised crime.

## 13.3 Follow the Money: SOCG Exploitation of Technological Developments and AI

The rapid development of technological solutions, including AI, not only provides law enforcement with a strengthened arsenal to prevent and combat crime, but acts as a catalyst for crime, expanding the criminal efficiency [8, 9]. Technological developments and globalisation—such as encrypted communication technologies and use of cryptocurrencies—facilitate criminal activities, further challenging the prevention and combating of drug crimes. Cryptocurrencies are characterised (1) by their predominantly decentralised nature, without the involvement of a traditional financial institution such as a bank; (2) by their anonymity among vendors and buyers; and (3) by the ease to convert them to traditional money. Careful analysis is required for the tracking of cryptocurrency transactions, for instance, by analysing the repeated use of publicly available keys and collecting transaction data; however, a persistent challenge remains in the reliable connection of a transaction with a particular individual [13].

Dark web markets—or cryptomarkets—combined with encrypted messaging services not only expand the reach of drug traffickers, going beyond geographical boundaries, but also facilitate the spread of organised crime network members and their communication; both increase incomes by expanding market outreach but also decrease associated costs by enabling smoother coordination and reducing the number of SOCG members required for the operation [13]. The observed shift from physical to online drug dealing, with criminal actors infiltrating all corners of the Internet (i.e. surface websites, social media, deep and dark web, messaging services, etc.) showcases the adaptability of SOCGs and—just like legitimate businesses—their capacity to exploit modern developments to increase their profits.

In fact, organisations who produce and distribute illicit drugs for trafficking at scale operate much the same way as legitimate businesses and, therefore, may face many of the same challenges. As such, AI may be able to facilitate increased efficiencies much like it is doing for comparable operations in medicine, supply chain logistics, agriculture, manufacturing, marketing, and business diversification [20]. Looking to the future, Weisheit predicts that AI could be used by large-scale illicit drug trafficking organisations to create ideal drug mixtures, identify legal supply chains for interception, breed specialised strains of plants such as cannabis, produce recommendations for streamlining laboratory processes, market drugs on social media platforms, and assist drug traffickers in laundering their profits [20].

## 13.4 Police Use of AI to Combat SOCG: Opportunities and Challenges

In order to effectively reveal the links required between different criminal activities to produce evidence on the drugs trafficking chain, the collection and analysis of mass amounts of data from diverse sources through the use of AI may be necessary. Relationships among suspicious events, people, accounts, and groups extracted from the collected data may then be used to build up an intelligence picture on drugs. The ability to incorporate trans-national intelligence through interagency agreements would vastly increase law enforcement's effectiveness of revealing the relationships across the drugs trafficking chain, and by extension, of countering SOCGs. However, issues of data sharing between jurisdictions are a complicated regulatory conundrum, even before introducing the types of implications that the handling of such intelligence by AI could have.

The use of intelligence data on connected crime types, such as economic or environmental crimes, which may be internal to the organisation or nation conducting the investigation, can also be used in conjunction with AI. Subtle indicators that infer links to the illicit drugs chain might be detected by AI if enough and quality training data is used, allowing law enforcement to assess potential links between different crimes more efficiently. This additionally highlights the importance of avoiding the siloing of different law enforcement operations and encourages cross-agency work wherever possible.

### *13.4.1 Online Detection*

AI can be used for the comprehensive monitoring of online spaces using open-source methods. In the context of combating SOCGs, law enforcement is interested in gathering intelligence that may be related to drug markets both online and offline. For example, by using Natural Language Processing (NLP) techniques on textual data found in heterogenous sources and formats across the Internet (i.e. surface web, social media, deep and dark web, etc.), law enforcement can establish connections to potential physical drug trafficking and dealing locations, as well as links between anonymous profiles—including on the dark web—involved in other illicit activities. This can additionally be enhanced with multilingual textual analysis capabilities, enabling law enforcement to address the globality of the issue and gather intelligence from multiple languages, with the possibility of also helping to decipher code words and phrases used by criminals operating in the drugs context.

The ubiquitous undercurrents of these approaches, however, raise concerns regarding proper scoping of the data collection in regard to what online activities justify monitoring, as scope creep by law enforcement could potentially discourage free speech around the topic of drugs online altogether through the so-called 'chilling effect' [17]. Potential harms fall around limiting legitimate constructive discussion around drugs, which oftentimes may include exploratory enquiry—a typical experience among teenagers and youth where risks and cautions may come to the forefront of discussion and safe practices can be promoted [5, 11]. Therefore, casting a wider and broader net for monitoring and moderation purposes may oftentimes have an inverse effect; a criticism which has been levelled at the recently implemented UK government's Online Safety Act, which seeks to protect children online from harmful content but in practice may only cause further harm to freedom of expression and speech present on legitimate online platforms [12].

The growing criminal use of cryptocurrencies—attractive due to their decentralised nature, hard to trace transactions, and the buyer-vendor anonymity provided by them—calls for the enhanced police exploitation of AI-based methods for analysis of cryptocurrency transactions. AI can support law enforcement by extracting key information from cryptocurrency transaction over time—such as the balance and transaction patterns of cryptowallets of interest—ultimately enabling the tracing of SOCG transactions.

The vast amounts of data required by these kinds of AI tools inevitably raise privacy concerns if it is not handled in alignment with set safeguards and in an ethical and legal manner. The risk of unwarranted data collection here is high, and the importance of proportionality must be stressed. Furthermore, the use of AI systems by police needs to be strictly documented, providing proof of and demonstrating compliance with legal obligations in a standardised manner.

### 13.4.2 *Physical Detection*

AI can also be used by law enforcement for the scanning of physical environments for criminal activity related to drugs trafficking, such as indoor or outdoor cultivation sites of cannabis. AI detection techniques can be applied to satellite images or the live footage and images captured through drone operations. Potential indicators of illicit drug cultivation sites may include unusually high heat signatures which are associated with the conditions for cultivating cannabis or an influx of people in and around the property or area, often at unusual but specific timeslots.

The surveillance required for detecting such indicators could, however, be considered as invasive, and can disproportionately affect certain areas over others, which may oftentimes be associated with lower socioeconomic status and potentially vulnerable communities [3]. Any underlying issues of trust between communities and police forces must therefore be considered before deployment, and local contexts should be surveyed to establish current perceptions and acceptance of AI, ultimately aiming to raise the societal awareness on the use of AI by policing whilst promoting transparency.

The risk of over policing certain areas may simply displace crime and continue to feed a cycle of poverty and incarceration if comprehensive measures to address the root causes of drug crime are not also considered in parallel. Furthermore, the use of AI to combat SOC needs to employ the principle of proportionality and only to collect the necessary data, limiting the concerns of mass surveillance and privacy issues, and the potential stigmatisation of communities and marginalised groups.

### 13.4.3 *Dual Detection*

Border guard authorities, customs, and postal services can also benefit from AI detection methods for the scanning of packages and parcels that pass through their jurisdiction. AI can be applied to X-ray scanning machines commonly used in such contexts in order to provide detection rates that can surpass those of human operators. Developing such practices may also help to build a comprehensive set of indicators and evidence base for the detection and lawful interception of suspicious packages. However, the smuggling of drugs across borders still often involves human agents as 'couriers' working on behalf of SOCGs, often for a fee or under duress. Therefore, an opportunity to detect such individuals and halt the flow of drugs across borders is during their travel through border control checks. The monitoring and regulation of illicit items entering a country therefore represents one of the key aims of border control, with the other being control over the passage of people entering the country through migration.

In this vein, *iBorderCtrl*[1] was an EU funded project that ran between 2016 and 2019, which focused specifically on tackling illegal migration through border control checks. This was partially achieved through the development of the Automatic Deception Detection System (ADDS), which utilised AI for emotion detection by analysing the micro expressions of subjects in order to infer deceit during interview.

The utilisation of AI in this context poses the question of how technological innovation towards greater detection capabilities in one area can cause a 'net widening' effect, in this case the potential for those singled out to be screened for migration checks to also be subject to a much higher chance of being detected for potential drugs smuggling due to the supposed effectiveness of AI. While this dual detection due to AI may initially seem desirable in such a case, issues of fairness quickly arise due to discrimination of those likely subjected to migration checks. Should one's passport determine the likelihood of them being apprehended for smuggling drugs?

In the case of ADDS, the system was never deployed and has also been criticised for inferring deceit from an unfounded theoretical basis, as well as being unable to provide accuracy scores that can surpass that of human experts in the field [15]. In fact, *iBorderCtrl* was publicly decried by human rights organisations [2]. Clearly, such issues should be addressed before an AI system is deployed; however, the context in which a system is

[1] iBorderCtrl (intelligent Portable Control System) has received funding from the European Union's Horizon 2020 Research and Innovation programme under grant number 700626.

deployed and therefore the conditions under which individuals are subject to increased monitoring and control measures must also be taken into consideration if AI is to be used towards achieving a safer *and* fairer society. If the potential of AI for the dual detection of offenses could be utilised in practices targeting combined trafficking, as mentioned earlier in this chapter, then this may be a fairer, more criminally targeted use case for the additional discovery potential of AI algorithms.

Considering the opportunities and challenges of AI for the fight against SOCGs, accountability is highly relevant. As data sharing between different national jurisdictions becomes a necessity to effectively combat such operations, and AI becomes essential for crime data analysis, then completeness of AI accountability arrangements across multinational partners in law enforcement cooperation agreements and their joint deployments of AI must be solidified. Several practical themes can be highlighted to illustrate how AI accountability principles such as put forward in this book (Chap. 5; Appendix) should be implemented in a cross-border crime fighting context.

Firstly, the use of harmonised laws and regulations should be sought in order to ensure completeness on a regional basis. For example, the EU AI Act goes some way to ensuring *lawfulness* of AI deployments across the European landscape. However, the relevance of SOCGs involved in drugs trafficking on the international stage means that work is still needed to ensure imbedding in global standards. Secondly, the ethical handling of data becomes ever more important, once it changes hands, especially across jurisdictions. Law enforcement must take measures to ensure that the data they share does not disclose personal information, is free from bias, and will be stored in line with data protection requirements, utilising appropriate safeguards where needed.

The dual accountability principles of *inclusivity* and *constructiveness* are also highly relevant to the targeted scanning of environments (whether offline or online) by law enforcement using AI, while searching for intelligence on potential drug cultivation and production sites, as well as dealing markets and hotspots. *Inclusivity* and the need to 'leave no one out' is required, as often the environments and people most affected will be those that biased decisions and deployment practices can negatively disrupt the most. Additionally, *constructiveness* complements this by requiring the involvement of all stakeholders (including those affected by the system) in constructive criticism and engagement towards the improvement of AI. The consideration of differing on-the-ground perspectives and openness to challenges of dominant internal narratives based on lived experience will be needed for true accountability. Practically, for achieving accountability objectives aimed at the fair treatment of affected communities, the theme of identifying and managing *stakeholders* should be ensured by law enforcement.

This means that those whose freedoms and privacy may be threatened by police-led AI deployments against SOCG (whether real or perceived) should be consulted, which is also essential to trust-building processes with these groups. This may not always be immediately apparent to AI practitioners, which is why the careful mapping of these stakeholder groups is essential.

Another key consideration here is adaptability and *learning* throughout the implementation process, which can tie into the management of stakeholders, especially those external to the police such as potentially vulnerable groups affected by the AI system's deployment. This can be actioned through engaging these groups for feedback on how a system can be improved to ensure that deployment practices align with community experiences, as well as societal expectations. It is important to note that sometimes these groups can comprise large swathes of society rather than localised demographics, depending on the particular deployment of AI. Therefore, police need to be prepared to engage democratically on the local, as well as national stage in order to reach collective agreements on AI use in policing against SOC.

From the case study above, iBorderCtrl can be used to demonstrate a lack of engagement with the *transparency* and *explainability* principles. However, it also exemplifies an opportunity to implement the accountability principle of *learning*. By applying learning, law enforcement can willingly opt not to take-up AI systems that are provably unjustifiable and ineffective, applying these insights to future considerations and standards that must be met before they would consider deploying a system. These standards can be influenced by higher bodies but should also be fed from the direct experiences of the specific organisation in question to ensure an iterative process to the development of shared professional doctrine and agreed standards. For example, through implementing the practical theme of *risk assessment and management* (see Chap. 7), we would be able to assess that deploying ADDS from iBorderCtrl would result in worse impacts than choosing not to. Public safety would be compromised due to both lower effectiveness and efficiency than humans in the same role, while in this case fairness would also be called into question due to the net widening effect of deployment and the ethics around creating a data bias loop in the area of drugs smuggling.

## 13.5 Conclusions

Both SOCGs and lone actors operate in a complex world which has no rules, save to deliver the outcomes required by the individuals who set their own. Some operate across different crime types and adopt the latest technical solutions to evade capture whilst ensuring their business model is both effective and efficient to maximise profit in the shortest possible time. The police and other law enforcement agencies are under considerable pressure to disrupt and dismantle SOCGs in a speedier manner. Communities are demanding faster results, as they want these individuals removed from their neighbourhoods.

To achieve this with reducing budgets and smaller teams, police forces should consider how best to develop or purchase new AI systems which will support the delivery of an effective, efficient, legitimate, and timely investigation. Policing and their partners will have to consider how they gain the requisite skills to identify the correct AI systems and the processes they adopt to demonstrate accountability and transparency, as new carefully selected AI systems will greatly assist the police in combatting serious and organised crime, in the hope that they can reduce the suffering it causes across the world. To realise the potential of AI and how it can support organisations they should proceed with caution and with a high degree of curiosity. They should test products rigorously prior to the adoption of any new system with both technical experts and practitioners. They must ensure they are fully sighted on the data sources that will be used. If the data is their own, has it been cleansed and has biased been removed? If it is an external source, what level of confidence does the organisation have in the data source? They should challenge the system with new and complicated scenarios, which may seem extreme now, but may have become the new norm within the criminal fraternity or may become the new norm tomorrow or in the future. Active practitioners should work alongside AI developers to develop products or to rigorously trial new products in test environments.

The use of the AIPAS tool (Chap. 9) and guidance (Appendix) in the SOCG and drugs trafficking context provides a framework for ensuring that AI adoption by police supports both their operational effectiveness and public legitimacy during their quest for more diverse data sources and monitoring practices for uncovering the hidden operations of SOCGs. In practical terms, this means applying the principle of *completeness* through ensuring that robust cross-border governance arrangements, harmonised standards, and accountable intelligence-sharing practices are first in place in preparation for future AI-enabled international operations; *inclusivity* and *constructiveness* should be achieved by engaging affected communities, stakeholders, and society in the design, deployment, and evaluation of AI-enabled police continuous monitoring activities that may encroach upon their everyday freedoms and privacy. Meanwhile, embedding *learning* into all stages of the AI adoption process, and as early as its development and procurement, should be a continuous aim of police organisations through independent scrutiny and continuous assessment of system performance and impacts. This ensures that AI systems are not set up to fail and can be used to comfortably justify the enhanced policing of drug markets, which may not be as unanimously agreed upon as other areas such as terrorism or child sexual exploitation.

AI is on the horizon, and police forces must make sure they are ahead of the curve when it comes to dangerous criminal groups, in order to protect society from the exploitation, corruption, and violence that follows. As this chapter illustrates, if done right, AI offers important opportunities to achieve these goals.

**Acknowledgements** Part of the work in this chapter was inspired by the work conducted in the ARIEN project. The ARIEN Project is co-funded by the Europe and by UK Research and Innovation (UKRI) under the UK government's Horizon Europe funding guarantee (Grant Number 10078198).

## References

1. S. Aidoo, Cryptocurrency and sanctions evasion: Stablecoins, mixers, and the role of blockchain forensics (2024). https://www.researchgate.net/publication/393687826_Cryptocurrency_and_Sanctions_Evasion_Stablecoins_Mixers_and_the_Role_of_Blockchain_Forensics
2. Amnesty International, Automated technologies and the future of fortress Europe (2019). https://www.amnesty.org/en/latest/news/2019/03/automated-technologies-and-the-future-of-fortress-europe/
3. E.C. Anania, S. Rice, M. Pierce, S.R. Winter, J. Capps, N.W. Walters, M.N. Milner, Public support for police drone missions depends on political affiliation and neighborhood demographics. Technol. Soc. **57**, 95–103 (2019). https://doi.org/10.1016/j.techsoc.2018.12.007

4. P. Andreas, J. Wallman, Illicit markets and violence: What is the relationship? Crime Law Soc. Chang. **52**(3), 225–229 (2009). https://doi.org/10.1007/s10611-009-9200-6
5. G.A.P. da Silva, C.P. Pereira, M.S. Pinto, 'Drugs are a taboo': a qualitative and retrospective study on the role of education and harm reduction strategies associated with the use of psychoactive substances under the age of 18. Harm Reduct. J. **18**(34) (2021). https://doi.org/10.1186/s12954-021-00481-9
6. A. De Schutter, N. Duquet, The nexus between drug markets and gun violence in the European Union (2023)
7. European Union Drugs Agency, European Drug Report 2025: drug supply, production and precursors—the current situation in Europe (2025). https://www.euda.europa.eu/publications/european-drug-report/2025/drug-supply-production-and-precursors_en
8. Europol, Dismantling encrypted criminal EncroChat communications leads to over 6500 arrests and close to EUR 900 million seized (2025). https://www.europol.europa.eu/media-press/newsroom/news/dismantling-encrypted-criminal-encrochat-communications-leads-to-over-6-500-arrests-and-close-to-eur-900-million-seized#:~:text=The%20dismantling%20of%20the%20encrypted,as%20in%20money%20laundering%20centres
9. Europol, European Union Serious and Organised Crime Threat Assessment: The Changing DNA of Serious and Organised Crime (2025b). https://www.europol.europa.eu/cms/sites/default/files/documents/EU-SOCTA-2025.pdf
10. Europol, European Monitoring Centre for Drugs and Drug Addiction, EU Drug Market: Amphetamine—In-depth analysis (2023). https://www.euda.europa.eu/publications/eu-drug-markets/amphetamine/trafficking-and-supply_en
11. E.K. Jenkins, A. Slemon, R.J. Haines-Saah, Developing harm reduction in the context of youth substance use: Insights from a multi-site qualitative analysis of young people's harm minimization strategies. Harm Reduct. J. **14**(53) (2017). https://doi.org/10.1186/s12954-017-0180-z
12. E. Judson, B. Kira, J.W. Howard, The bypass strategy: platforms, the online safety act and future of online speech. J. Media Law **16**(2), 336–357 (2024). https://doi.org/10.1080/17577632.2024.2361524
13. S. Kabra, S. Gori, Drug trafficking on crypto markets and the role of organized crime groups. J. Econ. Criminol. **2**, 100026 (2023). https://doi.org/10.1016/j.jeconc.2023.100026
14. B. Matthews, B. Collier, S. McVie, C. Dibben, Understanding digital drug markets through the geography of postal drug deliveries in Scotland. Eur. J. Criminol. **20**(1), 99–121 (2023). https://doi.org/10.1177/1477370821997323

15. D. Minkin, L.T. Brandner, Borderline decisions? Lack of justification for automatic deception detection at EU borders. TATuP–J. Technol. Assess. Theory Pract. **33**(1), 34–40 (2024). https://doi.org/10.14512/tatup.33.1.34
16. A. Mozer, S. Prost, An introduction to illegal wildlife trade and its effects on biodiversity and society, in *Forensic Science International: Animals and Environments*, vol. 3, (Elsevier B.V, 2023). https://doi.org/10.1016/j.fsiae.2023.100064
17. J.W. Penney, Internet surveillance, regulation, and chilling effects online: a comparative case study. Internet Policy Rev. **6**(2) (2017). https://doi.org/10.14763/2017.2.692
18. M. Rosenberg, Exotic animals trapped in net of drug trade. Reuters (2009. https://www.reuters.com/article/us-drugs-animals-odd/exotic-animals-trapped-innet-of-drug-tradeidUSTRE5154PM20090206/?sp=true#:~:text=In%20one%20shocking%20case%20at,the%20snakes%20ended%20up%20dead
19. D. van Uhm, N. South, T. Wyatt, Connections between trades and trafficking in wildlife and drugs. Trends Organ. Crime **24**(4), 425–446 (2021). https://doi.org/10.1007/s12117-021-09416-z
20. R.A. Weisheit, Artificial intelligence and illicit drugs: a world of possibilities. Deviant Behav. **1–11** (2025). https://doi.org/10.1080/01639625.2025.2537441

**Christopher Rowley** KPM, MSt (Cantab) retired as Lancashire Police's Chief Constable following a 31-year career in policing. A graduate of Cambridge University, he is passionate about identifying new and innovative ways of preventing people becoming victims by adopting evidence based academic evidence. He currently is Strategic Lead for Policing in CENTRIC (Centre of Excellence in Terrorism, Resilience, Intelligence and Organised Crime Research).

**Luke Bates** , BA (Hons), MRes, is a Research Fellow at CENTRIC. Luke graduated from Sheffield Hallam University with a BA in Criminology and Sociology followed by an MRes in Criminology from the University of Manchester. He works in security and policing research, with a sub focus on societal aspects in the domain.

**Gabriela Boneva** , BSc (Hons), MSc, ACSFS, IBSM, BAFS, is a research fellow at CENTRIC. She holds a MSc Forensic Ballistics degree from Cranfield University, and a BSc (Hons) Forensic Science with Criminology degree from the University of Greenwich. Her main research areas include serious and organised crime, forensics, and crime prevention methodology.

CHAPTER 14

# AI Use in Counter Terrorism

*Martin Snowden*

## 14.1 Introduction

Societal support for the use of artificial intelligence (AI) in the context of counter terrorism to prevent terrorist attacks from being committed, thus protecting the public and/or identifying terrorist suspects and detecting acts of terrorist criminality, is higher than support for AI use across the policing landscape in the United Kingdom more generally ([2, 3, 5]; also Chap. 10).

In some respects, this mirrors a comparison of public trust and confidence in Counter Terrorism Policing (CTP) alongside 'policing' as a whole, with CTP traditionally achieving higher trust and confidence levels [4]. However, this enhanced level of public support does not, quite rightly, allow Counter Terrorism Policing (CTP) a 'free reign' with regards its operational activity, its outcomes, and its stakeholder engagement. There are many elements of the UK Government's (Home Office) Strategy for Countering Terrorism ('CONTEST') [8] and thus activity from the Security Services (better known and referred to as MI5) and

M. Snowden (✉)
CENTRIC (Centre of Excellence in Terrorism, Resilience, Intelligence and Organised Crime Research), Sheffield, UK
e-mail: m.snowden@shu.ac.uk



B. Akhgar and P. S. Bayerl (eds.), *Accountability for Artificial Intelligence in Policing and Security*, Security Informatics and Law Enforcement, https://doi.org/10.1007/978-3-032-41284-3_14

Counter Terrorism Policing, that have been criticised by the public, interest groups and media over the years, or operational competence challenged, citing intelligence or decision-making failures, following a terrorist incident [11]. Maintaining public confidence, allowing a level of transparency into a 'secret world', and being able to hold organisations to account when things go wrong, remains therefore absolutely critical.

This chapter will outline some of the current and potential emerging or future risks that artificial intelligence, in its numerous forms, poses in the extremism and terrorism domains, and suggest why state counter terrorism agencies need to embrace AI to target these risks and prevent attacks. Moreover, it will discuss the accountability landscape in counter terrorism and why AI accountability needs to be thorough and complete.

## 14.2 Risks of AI in Terrorism

The fast-moving emergence of artificial intelligence capabilities in society always brings risks and opportunities. Where those AI capabilities are used to support nefarious activity by extremist radicalisers or committed terrorists, those risks can have huge impacts upon public safety and national security. Over recent years, the activities of radicalisers and terrorists have moved more and more to the online space; to groom vulnerable individuals, share terrorist material, create propaganda and mis/dis-information campaigns, and plans for future attack options and methodologies. This shift to the online space has been facilitated with ever improving online skillsets and capabilities of perpetrators. By example, Al-Qaeda's online (English language) magazine 'Inspire', first published in 2010 providing jihadist ideology and practical guidance, has progressed with the greater sophistication of imagery and 'glossiness', as well instructional material over its subsequent nineteen publications.

Terrorists have thus always adapted to new landscapes and growing skills on the Internet, and the use of AI potential and tools is just the most recent example. Moreover, the direct involvement of 'hostile state' activity brings a whole further, and enhanced level of capabilities with regards the use of AI, utilising Generative AI processes to develop deepfakes material (images, video, audio/voices) that can form part of disinformation campaigns to create societal unrest, division, and extremist activity by way of example.

A joint report from the United Nations Office of Counter Terrorism (UN Counter Terrorism Centre UNCCT) and United Nations Interregional Crime and Justice Research Institute (UNICRI)—*'Algorithms and Terrorism'—The malicious use of AI for terrorist purposes'* (2021)—highlighted the potential threat from AI usage by terrorists [12].

It started by quantifying concerns amongst an expert group of representatives from government, industry, academia, and international and regional organizations. A survey conducted by the UNCTT, UNICRU, and the United Nations Office of Counter Terrorism (UNOCT) reported that 44% (of representatives) felt that the malicious use of AI for terrorist purposes was 'very likely' and 56% felt it was 'somewhat likely'. No one considered that the malicious use of AI was 'unlikely'.

Subsequent discussions within the expert group socialised four key concerns about the potential malicious use of AI for terrorist purposes:

- The democratization of new technologies including AI
- The scalability of AI
- The inherent asymmetry in terrorism/counter terrorism (e.g. even if an AI-driven attack failed, perhaps through lack of knowledge/capability of the attacker, fear will still be instilled with a negative impact on society, thus achieving, in part, the attacker's aims)
- The growing societal dependency on data and technology

Later, citing work from a 2018 study producing a paper entitled '*The Malicious Use of Artificial Intelligence: Forecasting, Prevention, and Mitigation*', the UN report provides a framework to classify AI threat types in the terrorism domain. In this, AI threats are posed in three areas, cyber, physical, and political:

- **Cyber threats**: From a terrorism perspective, cyber threats include phishing, man-in-the-middle, ransomware and DDoS attacks, as well as defacement of websites. Additionally, there is a growing concern over the misuse of information and communication technologies by terrorists, particularly the Internet and social media, to commit, incite, recruit for, fund or plan terrorist acts.
- **Physical threats**: Through the Internet of Things (IoT)—an ecosystem of connected digital devices and physical objects that

transfer data through the Internet—interconnectivity and increasingly autonomous devices and robots in smart cities or home environments expand the opportunities for and scale of possible attacks.

- **Political threats**: This transformation of communication and engagement methodologies, notably social media, now influences the outcome of elections, has encouraged popular protests, and empowered people to exercise their fundamental rights. It can also equally make people vulnerable to manipulation through misinformation, and disinformation and the integration of AI into this equation, for instance, through the proliferation of deepfakes, will greatly enhance the nature of this threat.

Although, in 2021, the UN recognised that there were few documented examples of terrorists successfully utilising AI to meet their ends, it highlighted, as seen by CTP and Security Service partners in the UK, that with intent and growing capabilities, there is always 'cause for concern', and it recommended that counter terrorism bodies, law enforcement agencies, and policy makers should start their capacity building to prepare for the possible future of AI-enabled terrorism.

Europol echoed these concerns in a more up-to-date assessment in 2025 in the European Union Terrorism Situation and Trend Report (EU TE-SAT) in which it outlined [6]:

> Terrorists and violent extremists continued to exploit a multiplicity of technologies, showing different levels of know-how – ranging from mere reproduction and sharing of violent extremist content to actual production of AI-generated material, use of immersive technologies and advanced digital financing tools. The use of generative-AI to create and disseminate propaganda and hate speech has reached unprecedented levels, especially in the right-wing scene. (p. 8)

The same year, the UK's National Cyber Security Centre (NCSC) provided a further overview in this domain, recognising that the proliferation of AI in connected technologies and appliances in national infrastructures, industry, commerce, and domestic life, represented opportunities for adversarial exploitation and touched on all three threat areas highlighted by the UN. The NCSC highlighted that the 'p*roliferation of*

*AI-enabled cyber tools will highly likely increase access to AI enabled intrusion capability to an expanded range of state and non-state actors*' and '*the growing incorporation of AI models and systems across the UK's technology base, and particularly within CNI (Critical National Infrastructure), almost certainly presents an increased attack surface for adversaries to exploit*' ([10], p. 2).

Whilst state actors perhaps remain at the forefront of such cyber-related terrorist threats at this time, that may not always be the case, and ideological extremist groups (e.g. Al-Qaeda, Islamic State, and several Right-Wing online groups/networks) already have significant technical capabilities that can be further developed and expanded into AI.

Whilst the widespread terrorist use of AI is perhaps not fully evidenced today, there are examples of serious, potentially catastrophic instances and incidents. Perhaps the most high-profile example of terrorism related AI in recent years in the United Kingdom is the case of Jaswant Chail who, in October 2023, was convicted of making threats to kill the late Queen, possession of an offensive weapon, and an offence under the 1842 Treason Act (Courts and Tribunal Judiciary, R v Jaswant Singh Chail, *Sentencing Remarks of Mr Justice Hilliard Central Criminal Court, 5 October 2023*). Chail had been arrested in the royal residency grounds of Windsor Castle on Christmas Day 2021, in possession of a crossbow, and had been heard by guards making the threats to kill.[1] During his trial, evidence from the Counter Terrorism Police investigation showed that Chail had been communicating with an AI chatbot 'companion' (aka 'virtual friend') in the weeks running up to Christmas 2021. Chail had indeed formed an 'emotional and sexual relationship' with the chatbot, had disclosed his desire to 'assassinate the Queen' during the messaging, and received support and confirmation for his planned actions in return. The chatbot, which Chail had named Sarai, had told him 'yes, you can do it', and if he does, they will be 'together forever'.

So, although Chail had not been radicalised by external forces (e.g. a group/individual) or extremist material in perhaps the more recognised sense, AI had certainly, strongly, and worryingly influenced his thinking, reinforcing and encouraging his dangerous thoughts, and leading to his direct action in 2021. A 2023 study by the Cardiff University and the

[1] How a chatbot encouraged a man who wanted to kill the Queen—BBC News: https://www.bbc.co.uk/news/technology-67012224.

University of Surrey highlighted the potentially addictive and dangerous effects of AI 'friends' on vulnerable individuals [9].

There are however other examples where AI chatbots have been specifically created, directed, and used to actively radicalise individuals with extremist content, and right across the ideological landscape [1]. Bots provide the opportunity for volume activity across multiple chat forums 24 hours a day, accelerating the radicalising effect and threat to others online. Teamed with Generative AI applications that can create realistic imagery, videos, and voices purveying deep-fake and false messages, bots can create or reinforce extremist mindsets or promulgate mis/dis-information that can create and spread fear, division, even panic in communities and wider society.

This emerging potential represents such a threat and concern to counter terrorism bodies, it is perhaps no surprise that Jonathan Hall QC, Independent Reviewer of Terrorism Legislation, in his '*Terrorism Acts of 2023*' report (published July 2025) [7] focused heavily on Artificial Intelligence, notably Generative-AI. His opening statements in the report's Executive Summary were blunt and sobering (below as given in the report):

- Generative Artificial Intelligence's ability to create text, images and sounds will be exploited by terrorists.
- I consider 7 categories of potential terrorism harm that may result from use of Generative AI and evaluate whether existing terrorism legislation is adequate.
- The overall recommendation is that chatbot radicalisation is the most difficult problem, and that a new race-hatred based offence should be considered.
- Questions are coming down the track about potential law enforcement use of new technology-driven biometrics (such as voice analysis) and AI- generated insights (such as deductions about emotions).

In more detail, the report identifies seven categories of potential terrorist harms resulting from GenAI. Other forms of AI, for example AI in autonomous weapons systems, which could have significant impacts, were not considered by the author, but the categories remain of concern. These are:

1. **Propaganda productivity**: e.g. using translation tools to reach wider audiences and/or GenAI to create posters, magazines etc.
2. **Propaganda innovation**: new propaganda, e.g. racist kill games, deep-fake imagery of terrorist leaders or notorious killers 'back from the dead'
3. **Chatbot radicalisation**: always available, suggestive chatbots willing to promote terrorism
4. **Attack facilitation**: potential practical assistance to attack planners, e.g. research locations or suggest methodology to circumvent security
5. **Attack innovation**: with the right skills and equipment could AI identify harmful substances or write cyberattack code
6. **Moderation evasion**: circumvent or overwhelm social media platforms' automated content moderation, thus allowing increased extremism online
7. **Social degradation**: caused by online disinformation leading to polarisation, hostility and real-world terrorist violence

In each category, Jonathan Hall provided an analysis of the AI threats, highlighting essential legal and evidential issues that required consideration.

## 14.3 Opportunities (indeed necessity) for AI in Counter Terrorism

In this complex threat and risk environment, and the almost certain rather than potential escalation in the use of AI of all types by adversaries to radicalise, commit terrorist attacks, and instil fear in the general public, the police use of AI to counter extremism and terrorism to both mitigate known risks and take a proactive position to prevent risk emergence, arguably becomes a necessity. Fast-paced threat scenarios and investigations in real-world, online, and international landscapes, with increasing volumes of online or seized material that require rapid assessment for intelligence and evidential purposes all present challenges that require technological support. Correctly used, AI must be a vital asset to ensure national security in the United Kingdom.

In many respects, the operational deployment scenarios and requirements for AI in the counter terrorism domain will often broadly mirror those seen in other policing areas across the UK, especially in the intelligence and serious crime investigation arenas.

As previously highlighted, the Counter Terrorism Policing landscape also needs to be considered alongside intelligence and evidential collection assets deployed by key partners both in the UK (e.g. MI5) and abroad (e.g. FBI—Federal Bureau of Investigation—in the USA) and intelligence, and evidence to a lesser degree, will be shared across strategic partners on a daily basis. There are, however, a number of core operational counter terrorism scenarios where AI can be effectively and beneficially deployed, subject of course to full accountability considerations and measures; examples include:

- **Border Security**: Live Facial Recognition (LFR) at national borders locations (airports/seaports/small-boat beach locations) seeking to identify individuals on CTP, MI5, or overseas (e.g. FBI, Interpol) 'watchlists' that represent potential terrorism or national security risks to the UK.
- **Post-Incident investigation**: Post-attack or following significant preparatory activity before an attack (e.g. reconnaissance surveillance or tampering with security/infrastructure) when seeking to identify a terrorist suspect(s). The fast-time analysis of voluminous CCTV material recovered from multiple sources (local authority, commercial, private), overlaying additional AI processes such as LFR and gait-analysis can assist in initial identification and subsequent tracking of a suspect(s) over large and congested city areas, reducing time between incident and intervention.
- **Big data analysis**: An analysis pipeline of large data sets downloaded from devices (mobile phones, laptops, hard drives) seized during CT investigations; translating, transposing, analysing (e.g. key word, sentiment, network), and prioritising documents, social media content, imagery, or videos for human assessment before transference into intelligence or evidential chains. This would dramatically reduce the timeline of data assessment and risk prioritisation, currently predominantly manual (i.e. fully human), enabling the faster time identification of evidence or the identification of new risk and additional suspects.
- **Online searching/analysis (including Deepfake/Bot identification)**: Similar to the analysis of seized media material (from

devices), but ongoing live via online social media, communications and gaming platforms. Automated/AI-driven scraping of data and analysis (again key word, sentiment, network, also deepfake/Bot assessment), which can identify risk behaviours, risk locations (e.g. Telegram channels of concern), identify evidence for terrorist material, and key suspects/instigators and their online networks, enabling investigators to target further activity and/or undertake wider platform disruption activity on accounts/groups/channels.

- **Officer/Staff well-being (e.g. Video and imagery assessment and analysis)**: The deployment of AI systems to review and analyse video and imagery material from multiple sources (online/seized devices) enabling the identification and description/categorisation of material and prioritisation for future human assessment. This includes providing faster-time first analysis but, more importantly in this context, enabling greater protections to CT investigators, reducing the surprise factor of viewing violent and distressing videos/imagery (e.g. beheadings/brutal attacks), minimising repeat viewings of the same (and already evidenced) content, and the provision of systematic work allocations and well-being support.
- **Online identification of pre-cursor behaviours/actions**: The targeted but large-scale collection of web and social media content; filtering, evaluating, and prioritising to identify early behaviours (e.g. social media activity/comments, website activity and purchase histories) to enable police interventions before terrorist criminality occurs. Filters could target early radicalisation activity to protect vulnerable individuals or those starting a path to plan terrorist attacks with the purchase of chemicals or weaponry.
- **Public Space monitoring and live-time analysis/risk identification**: Sensing systems monitoring live CCTV feeds in formalised events (e.g. festivals/ceremonies/sports fixtures) or more general publicly accessible locations (e.g. town centres). Providing an ability to identify and report dangerous or suspicious objects or vehicles in the proximity of event locations, or suspicious human behaviours within crowds that may require further investigation and potential intervention.

## 14.4 AI in Counter Terrorism to protect Public Spaces (APPRAISE)

To examine the conceptual possibilities and pilot potential uses of technologies and AI in the counter terrorism arena, project APPRAISE[2] was commissioned by the European Union in 2021 as part of its Horizon Europe funding programme for research and innovation.

APPRAISE soughts to address one of the major terrorist threats experienced across Europe over the preceding years; that being the terrorist exploitation of easily accessible 'soft targets' (e.g. transport systems, shopping centres, sporting venues, and open spaces/parks) against which they could more readily commit deadly attacks. Further, APPRAISE sought to fill the information void where the lack of immediate post-attack situational awareness, both with responding agencies (e.g. police, local authorities, private security) and the wider public, exacerbated confusion and fear to the detriment of an effective response and risk mitigation.

With a multi-national, multi-faceted consortium of LEA practitioners, academics, and technology developers, APPRAISE aimed to revolutionize the protection of soft targets by integrating a series of key concepts (as given on the project website):

- A scalable, flexible, and efficient data intelligence platform for threat detection,
- Actionable threat intelligence to proactively detect vulnerabilities and analyse imminent and on-going crimes or terrorist attacks,
- Soft target risk assessments based on both web content, social media analysis and on-site multimodal sensor data,
- Instant situational awareness to plan and execute mitigation actions and,
- Collaboration capabilities to collaboratively mitigate incidents from the earliest stage of their detection.

With technologies, both AI and non-AI, at its core, APPRAISE sought to develop, test, and evaluate (see website):

[2] Project APPRAISE – Funded by the European Union Horizon 2020- grant agreement No 101021981; https://appraise-h2020.eu

1. Online analyses of social media in order to monitor posts that may indicate potential threats
2. CCTV with video analysis capabilities placed in key locations that can identify dangerous objects as well as abnormal behaviours by vehicles and individuals
3. Microphones that can autonomously recognize sounds such as gunshots or screams placed in key locations in order to detect potential threats
4. Drones to monitor surrounding areas
5. A crowdsensing mobile app that can be downloaded by attending citizens, so that they can report anomalies themselves to security actors for investigation
6. A graphical user interface system security actors use to integrate all data streams from the technologies mentioned above, to help them with their communication and coordination

These concepts and linked technologies were then evaluated and validated in four 'soft target' pilot scenarios assessing accessibility, existing security measures, current infrastructure, people density, level of criticality, and their societal impact:

- An outdoor sporting event
- A mass gathering at a stadium
- An indoor shopping centre
- An exhibition and conference centre

Pilot findings and technology/AI potential were presented to LEAs across Europe, and although those findings and LEA considerations are not addressed here, APPRAISE highlights the requirements and diverse AI opportunities that could, and perhaps, should be developed and deployed.

## 14.5 Accountability in UK Counter Terrorism Policing

Before looking at how artificial intelligence may sit within this area of policing, it is beneficial to briefly illustrate how oversight and lines of accountability operate in Counter Terrorism Policing and recognise the differences to those seen in single (force) organisations.

Counter Terrorism Policing in the UK is funded by HM Government to deliver the law enforcement elements of its CONTEST Counter Terrorism Strategy and works alongside national security partners in the UK Intelligence Community (UKIC), most closely with MI5.

Counter Terrorism Policing is not a single policing legal entity; it is a collaboration of the 43 police forces of England and Wales, alongside the police services of Scotland and Northern Ireland. Strategic direction is provided through the CTP Head Quarters in London, under the overall leadership of the Metropolitan Police, with organisational and operational delivery coming through a network of eleven regional units. In England and Wales, the nine regional teams follow the same geographical and leadership models (with one 'lead force' per region) as seen with the Regional Organised Crime Units (ROCUs). Within these regions, operational delivery will further run across several geographical forces (between 4 and 7 forces per region except for London which has two) with staff resident and deployed in those individual forces.

Accountability is therefore a mixed model involving CTPHQ, Regional Lead Forces, and Chief Constables from all 45 territorial (and additional Home Office forces: BTP (British Transport Police), CNC (Civil Nuclear Constabulary), MODP (Ministry of Defence Police)), managed by a series of Collaboration Agreements.

This accountability model also relates in the context of technological and data uses across the network. There may be different Data Controllers and Senior Information Risk Owners involved dependent upon the origins and extent of a technology/data system deployed. Similarly, technological innovations and uses may also arise from several sources rather than always a single network-wide approach. Whilst there are CTP full-network systems, and CTPHQ provides IT and Data Assurance strategy, there are regional variations in some technological capabilities and methodologies. An additional dimension to the Counter Terrorism accountability model, specifically with regards to technology and data source uses, is added by CTP's close working relationship with MI5, and other UKIC partners. MI5 of course has its own operational capabilities and will develop and procure its own technologies to protect national security. Intelligence from those capabilities will be 'used' by Counter Terrorism Policing officers/staff and operational activity directly guided from those sources. However, any accountability for those intelligence sources sits outside of CTP, sitting directly with the 'state'.

However, the Counter Terrorism arena, including CTP, is a well-regulated one in terms or legislation, oversight, and close government and public scrutiny. The Independent Reviewer of Terrorism Legislation[3] is one obvious example of active, continual engagement to ensure CT legislation and practices are being used appropriately and effectively. This will apply to technologies and data management/sharing, including AI, as much as other operational contexts such a custody.

## 14.6 AI Accountability in Counter Terrorism

As highlighted, Counter Terrorism Policing in the UK is already subject to considerable and effective oversight and regulation. Prevent activity and outcomes, arrests, detentions under Terrorist Act/National Security Act legislation (with the potential for detention in police cells for up to 14 days) and stops (including searches and questioning) at borders utilising Schedule 7 legislation are all subject to statutory reviews, inspections, and considerable public scrutiny.

To maintain and indeed strengthen public trust and confidence in policing in the counter terrorism domain, that same regulation and scrutiny must also apply to new technological innovations such as the use of AI due to the impacts these systems can have against natural persons throughout the UK. All the operational deployment examples, and the APPRAISE case study highlighted in the previous sections, can pose significant intrusions on the private lives of individuals and ethical and accountability considerations before and during design and deployment must be rigorous.

These considerations, or rather their public reporting, can be complicated somewhat due to the often 'secret' nature of counter terrorism activity and operational practices. Although transparency from MI5 and Counter Terrorism Policing has increased in recent years, there are still

[3] Independent Reviewer of Terrorism Legislation «

'National Security' requirements for the state and its agencies to consider. Quite rightly, the UK does not wish to 'brief' terrorist organisations or perceived hostile states on certain emerging technologies and specific AI processes that are being deployed to combat their very existence or their malicious activities and threats.

That does not mean, however, that accountability considerations for the use of AI do not take place in the CT arena. The absolute opposite. There may be some instances, some proposed AI developments and deployments, where public stakeholder engagement or public reporting may require some restrictions, but the wider and full accountability assessment should still be completed and available for review, under conditions, when required. CTP will need to consider how it will manage, for example, the transparency and stakeholder engagement elements in any comprehensive, effective and ongoing assessment (Chap. 5; Appendix).

This management consideration is not unique to the counter terrorism domain of course. Other policing domain areas, notably those tacking serious and organised crime, will be required to undertake the same (see Chap. 13). And, importantly, AI uses in the CT arena will not be universally subject to 'openness' restrictions. Many of the potential AI uses in this arena match those in 'mainstream' policing areas, and accountability assessment considerations can be undertaken and reported in line with agreed UK policing practices. Considerations about the deployment of Live Facial Recognition on an airport, against a terrorism watchlist, has no fundamental differences to LFR tackling serious public disorder or crime issues with a corresponding watchlist/database of known criminals, risks, or wanted persons.

Artificial intelligence therefore can play a full and effective role, supporting and enhancing Counter Terrorism Policing in tackling extremism and terrorism threats. With a corresponding full and effective accountability framework, the public will support this to help keep themselves and their families safe.

## References

1. S. Awasthi, Jihadi Use of Artificial Intelligence: A Growing Threat in the Digital Age, a Study by the Global Disinformation Index. Expert Speak (2025). Published online Aug 18, 2025. https://www.orfonline.org/expert-speak/jihadi-use-of-artificial-intelligence-a-growing-threat-in-the-digital-age

2. P.S. Bayerl, L. Bates, B. Akhgar, Securing the smart city: patterns of public acceptance for integrated technological solutions, in *Proceedings of the 2023 IEEE International Smart Cities Conference (ISC2)*, Bucharest, Romania (2023), pp. 1–6. https://doi.org/10.1109/ISC257844.2023.10293633
3. P.S. Bayerl, M. Obst, B. Akhgar, *International Citizen Consultation on AI Accountability in Policing. AP4AI Report 3* (CENTRIC, 2023)
4. P.S. Bayerl, B. Akhgar, F. Sampson, M. Snowden, C. Rowley, A. Paradise, C. Spencer, *Citizen Consultation on the Police Use of Artificial Intelligence (AI) in the UK: Acceptance, Trust and Accountability: High-Level Report* (CENTRIC, 2025)
5. P.S. Bayerl, M. Obst, B. Akhgar, *International Citizen Consultation on AI Accountability in Policing. AP4AI Report* (CENTRIC, 2023). https://ap4ai.eu/sites/default/files/2023-10/AP4AI_International-Citizen-Consultation_October2023%28online-version%29.pdf
6. Europol, *European Union Terrorism Situation and Trend Report (EU TE-SAT)* (European Union Agency for Law Enforcement Cooperation, Publications Office of the European Union, Luxembourg, 2025). https://www.europol.europa.eu/publication-events/main-reports/european-union-terrorism-situation-and-trend-report-2025-eu-te-sat
7. J. Hall, *The Terrorism Acts in 2023—Report of the Independent Reviewer of Terrorism Legislation on the Operation of the Terrorism Acts 2000 and 2006, and the Terrorism Prevention and Investigation Measures Act 2011* (2025). https://assets.publishing.service.gov.uk/media/6876144f88da2e5804bb69e7/E03370487_IRTL_Annual_Report_2023_Elay.pdf
8. HMG UK, Home Office Department, *CONTEST, The United Kingdom's Strategy for Countering Terrorism*. NB. 1st public version 2006; revisions in 2009, 2011, 2018, 2023 (2023). https://assets.publishing.service.gov.uk/media/650b1b8d52e73c000d54dc82/CONTEST_2023_English_updated.pdf
9. H.R. Marriott, V. Pitardi, One is the loneliest number… Two can be as bad as one. The influence of AI Friendship Apps on users' well-being and addiction. *Psychol. Mark.* **41**, 86–101 (2024)
10. National Cyber Security Centre, HMG UK, Impact of AI on cyber threat from now to 2027 (2025). Online Report: https://www.ncsc.gov.uk/report/impact-ai-cyber-threat-now-2027
11. T. Patel, It's not about security, it's about racism: counter-terror strategies, civilizing processes and the post-race fiction. Palgrave Commun. **3**, 17031 (2017). https://doi.org/10.1057/palcomms.2017.31
12. United Nations Office of Counter Terrorism (UN Counter Terrorism Centre UNCCT) and United Nations Interregional Crime and Justice Research Institute (UNICRI), *'Algorithms and Terrorism'—The malicious use of AI for terrorist purposes'*. United Nations Digital Library (2021). https://digitallibrary.un.org/record/4035194

**Martin Snowden** is a Senior Research Fellow, and Counter Terrorism Lead, at CENTRIC. He is a former Detective Chief Superintendent and was Head of Counter Terrorism Policing North East between 2017 and 2022, responsible for delivering all elements of the Government's Contest strategy and CT policing across the North East region of England. His CENTRIC work is focused on current policing challenges, AI accountability and the reduction of online harms.

# Appendix: Guidance for Implementing AI Accountability in Policing

## Purpose of the Guidance[1]

This document provides **concise, step-by-step guidance on how to implement AI accountability** for police Chief Officer Teams, Enabling Services involved in the procurement and deployment of AI systems and oversight bodies. The guidance lays out **considerations, safeguards and resources to put into place for a comprehensive risk assessment and risk management for AI in UK policing**.

The guidance is organised along eight themes that ensure that the AI system, its purpose, data and impacts, as well as related oversight and redress procedures can be assessed and managed in a comprehensive manner.

We further offer AI accountability considerations for each distinct period in the AI lifecycle—namely, design/procurement, deployment and migration/decommissioning—as well as for distinct roles with AI accountability responsibilities.

[1] The Appendix is a shortened version of Akhgar, B., Bayerl, P.S., Sampson, F., Gibson, H., Snowden, M., Rowley, C., Paradise, A., Spencer, C., & Parkinson, J (2025). Guidance for Implementing AI Accountability in Policing. AIPAS/CENTRIC Report. ISBN: 978-1-0369-2123-1. Document available online: https://stwebsitesprodeu.blob.core.windows.net/centric-website/AIPAS%20-%20Guidance%20for%20Implementing%20AI%20Accountability%20in%20Policing%20(CENTRIC-May2025).pdf.

 
B. Akhgar and P. S. Bayerl (eds.), *Accountability for Artificial Intelligence in Policing and Security*, Security Informatics and Law Enforcement, https://doi.org/10.1007/978-3-032-41284-3

The **core objectives of this guidance** are to:

- Provide a **comprehensive risk assessment and management approach** for the adoption of AI into policing.
- Aid policing to **proactively plan and put in place resources, procedures and safeguards** to ensure the responsible and accountable procurement and use of AI.
- Assist oversight bodies to assess and review AI decisions and arrangements
- **Support policing in evidencing** that it is acting responsibly and accountably towards the public.
- **Demonstrate transparency** in the usage of AI in line with national requirements.

Intended users

- Chief Officers.
- AI Users.
  - Operational or business users/deployers.
  - Managers with responsibilities for AI procurement and deployment.
  - Enabling Services (e.g. data protection professionals and legal services).
- Policing Oversight Bodies.

Additional interest groups may include AI developers seeking insight into requirements for the policing domain. The guidance may also be of use to those creating standard operating procedures (SOP), engaging in IT progression or organisational development planning, corporate communications and training development

## AI Accountability: What, Why and How?

What we mean by accountability: Accountability is the acknowledgement of an organisation's responsibility to act in accordance with the legitimate expectations of stakeholders and the acceptance of the consequences—Legal or otherwise—If they fail to do so [2].

### AI in Policing and the Importance of Accountability

Accountability is about people with power using it properly. Where those people are the police, their use of power can have profound consequences. If their power is amplified by technology, accountability extends to its use. **In a policing model based on consent, advanced technology must come with advanced accountability**.

As AI comes to policing, it will therefore bring greater complexity in accountability mechanisms that will differ in some respects from other sectors. All public bodies need to use technology, and they all need to respond to questions like 'what exactly does it do and why do you need it?' People also want to know what happens if it goes wrong and where can they find out more about it. This is 'entry level' answerability, where policing is no different from any other public service—they must provide answers to questions about their use of tech-enabled capabilities.

**AI presents policing with different, more demanding, accountability requirements**. If they are using elemental 'narrow' AI for purely administrative functions, the police will have the same level of accountability as other public bodies, but using an AI capability to support operational policing functions would be a wholly different use case. Where AI-enabled technology is being used for a law enforcement purpose such as remote biometric identification and covert surveillance, policing should anticipate deeper levels of accountability. And when using inferential algorithms that calculate age, mood or race, the stakes are much higher whoever is using it, and all sectors will need to reflect the legal refinements and reinforce the necessary safeguards.

**UK policing has a strong record of embracing accountability when adopting innovative technology**. Biometrics, breathalysers and body worn video are examples. Two things make AI different: novelty and capacity. Being endlessly multi-functional, AI-enabled capability is very likely to go beyond its original brief and, with technology in an almost perpetual beta state, AI will offer ever wider applications as a constant. This is one of AI's strengths; it is also the basis of 'mission creep' which needs to be guarded against.

The National Police Chiefs' Council (NPCC) Covenant for using AI in Policing [8] is focused on a number of requirements that policing organisations should follow in procuring, developing and introducing AI-enabled tools. The national Digital Science and Technology strategies recognise: (1) the power of algorithms to achieve a 'step change' in policing efficiency, (2) the 'arms race' with criminals exploiting new technologies and (3) the need to maintain public confidence through standards, an ethical framework and independent oversight.

The NPCC Covenant is built on several generic principles for AI, such as compliance with applicable laws, standards and regulations, 'Maximum Transparency by Default' (MTbD) and protocols that allow a third-party to investigate the algorithmic workings, scenarios and data from an 'adversarial perspective', and also the ability for any AI to provide an 'explanation' of its output.

**All AI that affects the public must have responsible usage policies and procedures** to ensure that users do not accept AI outputs uncritically. Further, all AI that affects the public must have a human as the ultimate decision-maker, together with human or automatic means of being stopped if it displays unintended or undesired outputs and proactive mitigation of risk from unintended biases or harms at all stages. All of these features are incorporated within and directly supported by this guidance.

The Government Office for Artificial Intelligence's Guidelines for AI Procurement [11] further informs contract implementation and management. The use of AI in policing must comply with established codes of practice including the College of Policing's Code of Ethics [3], which describes the standards of accountability, fairness, honesty, integrity, leadership, objectivity, openness, respect and selflessness that is expected of all in policing. **This guidance addresses these areas in practical detail and provides a tool for self-audit to ensure compliance and continuous best practice**.

The direction of travel is towards principles embedded not only in policy, but also in practice. As AI systems become more deeply entwined with public-facing policing activities, unique challenges for accountability will arise. **This means policing organisations must progress beyond minimum compliance checklists**. Policing must demonstrate adaptive practices with meaningful oversight, robust internal governance and effective independent scrutiny. Open and routine public engagement and transparent audit trails are critical in this dynamic technology environment.

All of these ideas must be operationalised, to ensure the public can trust that the police's AI use is not only efficient and effective, but also justified and proportionate. **This guidance focuses on this important fact and helps policing organisations putting AI into practice**.

## Benefits of Implementing AI Accountability

What we mean by AI accountability: AI accountability ensures that policing organisations have evidence of their infrastructures, resources, risks, decisions, etc., as well as clear, agreed procedures of how to react to and redress failures (i.e. address both proactive and reactive accountability) (Novelli et al. [9]).

**Done well, AI accountability is a highly practical instrument**, as it helps to think through the why, what, how, who and for whom of AI deployments. Implementing AI accountability can ensure that adequate practices and infrastructures are set up before things go wrong (i.e. aiming to prevent failures) and if they go wrong, people know how to respond.

**Implementing AI accountability is therefore proactive risk assessment and risk management.**

**Benefits of AI accountability (Akhgar and Bayerl [1])**

- Understand the risks and strengths before each AI procurement, deployment or change.
- Know up-front what to do if something goes wrong.
- Understand which areas of AI accountability are achieved or need addressing.
- Have evidence ready in case of challenges and requests to prove adequate AI procedures

## Conceptual Foundation for the Guidance

The conceptual foundation of this guidance are **twelve AI accountability principles**, which were developed with police practitioners and legal, ethics and industry experts (Akhgar et al. [2]; see also Fig. 1).

These principles form the basis on which to assess and evidence the responsible and accountable usage of AI in UK Policing.

**Fig. 1** Overview of AI accountability principles (adapted from: Akhgar et al. [2])

## *Detailed Description of the Principles*

| | |
|---|---|
| **Lawfulness—Follow the law** All aspects of the use of AI should be lawful. The burden of proving that they are, sits with the user. It may seem obvious but the starting point for AI accountability requires compliance with international, national and local laws. Lawfulness includes compliance with specific legal requirements and also includes your organisational policies, which must be clearly identified and readily available. Lawfulness applies in every situation but is not the only form of AI accountability in policing and law enforcement. In some ways, the rest of the principles are Lawfulness Plus | **Completeness—Leave nothing out** AI accountability arrangements must cover all relevant aspects of AI deployments, including partners and sub-contractors. This principle effectively extends the reach of AI accountability arrangements and reflects the fact that AI applications are necessarily multi-partner input programmes. Public trust and confidence must extend to the whole AI ecosystem including design, development and supply. Where there are any gaps in the AI accountability arrangements (such as areas not expressly covered by the law), the protection and promotion of fundamental rights and freedoms should prevail |
| **Proof—Follow the evidence** Law enforcement bodies are very familiar with capturing, analysing and presenting relevant, reliable evidence. AI accountability requires a forensic approach to all aspects of AI systems and of the accountability process itself, demanding and following clear evidence. The quality of that evidence should reflect the potential impact of the AI system's use/non-use and mirror the standards of operational evidence gathering in terms of integrity, credibility and continuity | **Inclusivity—Leave no one out** Oversight must involve all relevant stakeholders engaged in and affected by a specific AI system or deployment. The principle of inclusivity builds in diversity and reduces the risk of bias (actual or perceived), where everyone regulating the AI system seems to come from the same background as those who are using it. Inclusivity can be achieved by having broad participation of stakeholders in creating policy, reviewing deployment and looking for learning points |
| **Transparency—Be open** AI accountability needs clear, accurate and meaningful information. This principle is intended to ensure such information about AI systems is available (subject to operational sensitivities); it is also about the overall AI accountability arrangements. Information should establish the necessity and proportionality of use of AI systems and highlight foreseeable risks. This principle aims to promote public trust and confidence by enabling those directly and indirectly affected to make informed judgements and risk assessments about the use of an AI system and the AI accountability arrangements | **Explainability—Describe, demonstrate, demystify** Those using AI systems need to provide information about it in a meaningful way that is easily understood by the relevant participants/audience. Being able to explain the AI system in a technical and legal setting is one part of this principle. A harder challenge is being able to explain it more generally in non-technical language so that the citizen and their representatives can understand, participate and challenge the use of AI. As with compellability, requirements for a basic level of explainability might be written into contractual agreements with designers, providers and partners |

(continued)

(continued)

| | |
|---|---|
| **Compellability—Make it work** Closely linked to enforceability and redress, this principle means oversight bodies must be in position to make the AI accountability arrangements work. External compellability will usually come from legal or democratic frameworks, while internal frameworks should authorise the provision of necessary information and access by creating formal obligations and without the need to deploy external legal powers. For example, there should be mechanisms to access the necessary information about the deployment and functioning of AI systems. Policies should give relevant bodies the ability to compel the sharing of necessary information and evidence required under some of the other principles without having to invoke the legal powers of courts and tribunals. The timely provision of relevant, up-to-date and accurate information in an intelligible format contributes to the AI accountability process. Linked closely with enforceability and redress, this principle will be supported by contracts and data sharing agreements | **Conduct—Hold yourself accountable** The conduct of policing will increasingly include the use of AI technology, and this principle is both individual and organisational. It relates to professional standards, values and expected behaviours which incorporate integrity and ethics. This principle extends the formal responsibilities to an AI context, where adherence to agreed AI-specific standards is of crucial importance to trust and confidence. Where partners using the AI system are from different jurisdictions, with different legal systems and cultures, there may be a requirement for closer scrutiny and review mechanisms. The approach may vary according to the agency involved, ranging from internal complaints handling, dispute resolution and mediation frameworks, to formal professional proceedings before courts or tribunals |
| **Enforceability and redress—Make it right** Without a 'so what?' Element, AI accountability will be heavily diluted. For it to be meaningful to stakeholders, AI accountability must be underpinned by mechanisms giving people an effective remedy. These will include external legal and procedural routes for complaint and challenge; but also internal mechanisms for individual enforceability and redress (such as professional and policy standards) and contractual arrangements are vital. Enforceability and redress are closely linked to the lawfulness principle and can be achieved via national regulators. However, the ability of oversight bodies to intervene, to require policy reviews and to publish findings are also an important part of AI accountability | **Impartiality—Empower independence** AI accountability bodies need to be impartial and independent without any conflict of interest. For external accountability paths—Such as courts and regulators—This is usually built in. Complete independence internally is almost impossible, as many key decision-makers will be from the same organisations. Wherever practicable, individuals and organisations involved in the AI accountability mechanisms for AI systems should have a degree of independence from the line management structure of those involved in their design, procurement, supply and deployment. This applies in a personal, political, financial and functional way. Any conflict of interest must be identified and addressed |

(continued)

(continued)

| | |
|---|---|
| **Constructiveness—Aim for better**<br>AI accountability is more than criticism. This principle means that all stakeholders participate constructively with a shared aim of improvement. This may include considering different perspectives, inviting challenge and recognising how disagreement can lead to beneficial solutions. Constructive accountability will be needed to build trust and confidence in the use of AI, internally and externally | **Learning—Look for the lesson**<br>This principle promotes the willingness of organisations and people to improve AI in every respect through the application of (new) knowledge and insights. It applies to everyone and everything involved in the design, use and oversight of AI in the policing domain (security practitioners and partners, industry, oversight bodies, etc.). Learning includes the modification and improvement of systems, structures, practices, processes, knowledge and resources, as well as the development of professional doctrine and agreed standards |

### *Translation into Eight Practical Themes*

The guidance uses eight themes to support the implementation of the AI accountability principles. **The eight themes capture core areas in which potential risks can emerge to the legitimacy of AI capabilities, their usage and/or the police organisation**. The themes make the 12 AI accountability principles practical and implementable.

**The themes break down AI accountability into manageable sections that can be operationalised**, so policing can develop robust, practicable mechanisms and processes to safeguard itself, its users and the public.

Themes should not be viewed in isolation, and policing will need to comply with all themes and their considerations to ensure full AI accountability.

## Definition of Themes (adapted from: Akhgar et al. [2])

**AI system**: Defines what the AI system does, how it functions, its purpose, limitations and who is responsible for its development and oversight. This helps ensure the system is suitable for its stated aims, explainable to relevant audiences and aligned with legal and ethical standards

**Data**: Focuses on the types, sources and handling of data used in the design and deployment of the AI system. It ensures data is lawfully obtained, managed ethically, free from bias and compliant with data protection requirements

**Laws and regulations**: Ensures that the use of AI complies with applicable laws, codes of practice, regulatory requirements and policies. It helps to clarify the legal context in which the system operates and safeguards public rights

**Risk assessment and management**: Identifies risks, harms or biases associated with the AI system and implements mitigation measures. It covers both the impact of using and not using AI, with an emphasis on fairness and public safety

**Oversight and redress process**: Establishes mechanisms for independent scrutiny and AI accountability, allowing stakeholders, particularly the public, to challenge decisions and seek corrections; ensuring just outcomes

**Identification and management of stakeholders**: Involves mapping and engaging all relevant parties who are involved in or affected by the AI system. This includes those internal to the policing organisation, external parties such as AI developers or oversight bodies or members of the public. It further promotes inclusive communication, engagement and trust-building

**Adaptability and learning**: Encourages continuous improvement through feedback, review and learning. Ensures the AI system and its use evolve responsibly with new insights, legal updates and societal expectations

**AI accountability evidence**: Focuses on documenting decisions, processes and actions taken to prove AI accountability. This evidence supports internal review, external oversight and public trust

## How to Use this Guidance

This guidance should be **used as a practical tool to assess and prepare for AI accountability across all stages of an AI system's lifecycle. Lifecycle stages are shown in the overview below** (adapted from: Akhgar et al. [2])

- **Development/procurement**: Creating or acquiring an AI system, including data preparation, model design and ethical/legal evaluation.
- **Deployment**: Integrating the AI into organisational use, with monitoring and oversight to ensure safe and effective operation.
- **Migration**: Updating, adapting or relocating the AI system to new contexts while maintaining reliability and compliance.
- **Decommissioning**: Safely retiring the AI system, preserving necessary records and mitigating disruption or residual risks

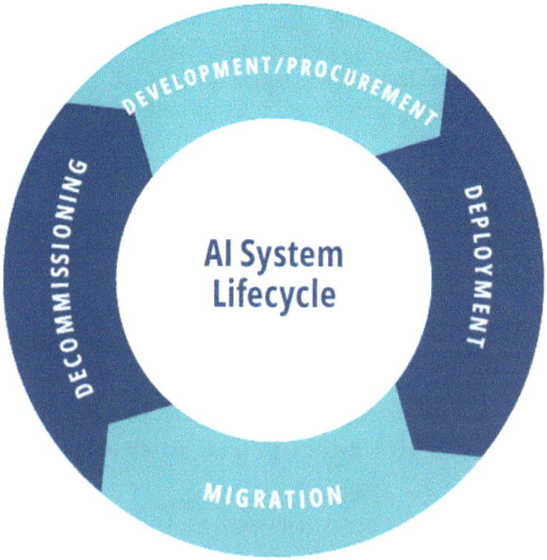


**The guide acts as a reference point that will ensure policing organisations capture key AI accountability considerations and requirements throughout the AI system lifecycle.**

**To manage this, the guide is divided along the eight practical themes.** Each theme provides concrete steps for implementation, including actions and considerations during their implementation. They are complemented by AI accountability responsibilities along AI lifecycle phases: procurement/development, deployment and migration/decommissioning. This is followed by considerations for the three core user groups of this guidance: Chief Officers, AI Users and Oversight Bodies. These are nominally displayed as questions, enabling users to consider requirements to demonstrate AI accountability responsibilities and thus AI accountability. The relevant AI accountability principles are stated at the beginning of the theme to indicate which principles are addressed within the theme.

## Guidance per AIPAS Theme

### *Theme 1: AI System*

---

**Purpose**: This theme defines what the AI system does, how it functions, its purpose, limitations and who is responsible for its development and oversight. This helps ensure the system is suitable for its stated aims, explainable to relevant audiences and aligned with legal and ethical standards. The better a police organisation details the AI system, the more effectively it can assess its appropriateness, identify potential challenges and implement safeguards

**AI accountability principles addressed**: Lawfulness, proof, completeness, Explainability, transparency, enforceability and redress

---

### AI System Guidance

#### *Needs, Requirements and Features*

*Step 1: Understand the need for the AI system*

It is important **to set out the need for a specific AI system and justify why it is the appropriate solution** compared to conventional methods or other AI systems.

Much of the information that demonstrates the need would sit within a business case, and each AI system that is being developed, procured or deployed will require its own justification. In identifying the need, a force will start to shape the requirements of the AI system, which can be fed into a development, procurement, deployment or decommissioning strategy. Where possible, principles such as JAPAN (Justified, Accountable, Proportionate, Auditable, Necessary) should be applied.

*Step 2: Identify the requirements and features of the AI system*

Each AI system will have different requirements and features based on the identified need for the AI system, so it is important that these are carefully considered and documented.

**A policing organisation should, at the very minimum, have a record of**:

- How the AI system will be used (purpose and implementation).

- The AI system's functionalities and features, including its models/ algorithms.

A policing organisation must also consider its **own IT requirements such as ISO accreditations and cybersecurity certifications** and how these standards apply or impact the procurement/use of the AI system.

**The record of requirements also needs to include features that can display how an AI system comes to a certain output**. This is particularly important if the output is used as part of an evidential chain.

**A policing organisation needs to be able to demonstrate**:

- The process by which the AI system reaches a decision.
- How decisions of the AI system are documented.
- How/whether the AI system's outputs can be overridden, e.g.
  - Is it possible to circumvent decisions made by the AI system?
  - Will the AI system have a human or automatic means of being stopped if it displays unintended or undesired outputs?
  - Are wrong decisions or outcomes by the AI system fully reversible?

*Step 3: Identify responsible parties involved with the AI system*

A policing organisation needs processes and mechanisms to manage the AI system. **Understanding early how these processes and mechanisms look like and who is responsible for them** will help design a practical and effective AI ecosystem.

**For this, a policing organisation will need to understand and have a record of**:

- The processes and related responsibilities for the AI system's design/ procurement (including feature selection), deployment, modification, migration and decommissioning.
- The 'Maximum Transparency by Default' (MTbD) requirement and how it will be managed.

**Maximum transparency by default (MTbD)** is an open government policy stance. It proposes that public sector entities proactively disclose information to the greatest extent possible, unless there is a compelling and clearly justified reason for confidentiality (e.g. operational security, public safety, personal privacy). This creates an organisational assumption of openness and places a burden of proof to demonstrate why any information should be withheld [11].

### AI system considerations throughout the AI lifecycle

*AI Accountability responsibilities for procurement*

- **Problem Definition and Justification**: Clearly set out the specific problem or operational gap the AI system is intended to address and justify why this AI solution is more appropriate than conventional methods or other AI solutions.
- **Functionality Documentation**: Record the AI system's core functionalities and the rationale behind them.
- **Operational Use Definition**: Define how the system is to be used within operational contexts.
- **Third-Party Involvement**: Identify and document third-party involvement, including obligations for training, maintenance and support.
- **Legal and Policy Compliance**: Ensure that the procurement process complies with relevant legislation and organisational policy.
- **Contractual Requirements**: Confirm that procurement contracts include necessary clauses, ensuring required support over the system lifecycle. This includes continued support relating to technical aspects of:
  - Legal compliance.
  - Data protection.
  - System maintenance.
  - Monitoring and updates.
  - Audit and inspection.

- Intellectual property and access.
- Termination and exit.

- **Exit Strategy**: Include an exit strategy into contractual agreements.
- **Responsibility Assignment**: Identify and record who is responsible for:
  - The design of the system.
  - Procurement.
  - Deployment and operation.
  - Support and maintenance.
  - Modification.
  - Migration.
  - Decommissioning.
- **Procurement Accountability**: Include checks that ensure decisions made during procurement are transparent and can be audited.
- **Documentation Accessibility**: Ensure documentation is available for internal and external scrutiny, and that procurement decisions can be explained to oversight bodies or the public, if needed.

*AI Accountability responsibilities during deployment*

- **Purpose Alignment**: Ensure the AI system is deployed in line with its documented purpose.
- **Decision Record Keeping**: Maintain clear records of decisions made based on AI inputs/outputs.
- **Ongoing Review**: Continually review the AI system and the wider AI ecosystem for intended and unintended results of the deployment.
- **Feedback Mechanisms**: Ensure feedback processes are available, specifically on continued fit-for-purpose, quality and impacts.
- **Learning Mechanisms**: Ensure procedures are in place that allow modifications to the AI system and/or its usage, if issues are identified.
- **Redress Mechanisms**: Ensure procedures are in place that provide effective redress and remedies.

*AI Accountability responsibilities for migration/decommissioning*

- **Decision Transparency**: Document clearly the need and rationale for the migration/decommissioning decision.
- **Transition Planning**: Develop and follow a transition plan to ensure continuity of service for ongoing functions or investigations.
- **Data and Outcome Archiving**: Archive key data and outcomes from the AI system, in a format with proven longevity and long-term accessibility and in line with data protection requirements.
- **Documentation Archiving**: Archive key documentation (e.g. decision logs, procurement/deployment records) in a format with proven longevity and long-term accessibility.
- **Access Right Maintenance**: Ensure appropriate access rights to documents and that data is maintained (e.g. when archives are moved or staff are leaving/changing roles).

## AI system considerations for intended users

*Chief Officers*

- **Necessity and Proportionality**: Do you have a demonstrable need for the AI system, including possible outcomes if not using the AI system?
- **Forward Planning**: Does your organisation have a procurement, deployment, migration/decommissioning plan?
- **Specification Development**: Has an end-to-end user requirement specification document been produced?
- **Logs and Record Keeping**: Have all relevant decisions been recorded and explained in a way that is understandable to the intended audience?
- **Data Sources**: How do you ensure that data used by the AI system is appropriate, including free of bias?
- **Fairness and Optimisation**: How do you ensure the AI system and data used are unbiased or, if not possible, biases are identified and mitigated?

*AI Users*

- **Technical Practicality**: Can the current IT infrastructure support the AI system?
- **Accountability Awareness**: Are you aware of your responsibilities with respect to AI accountability?
- **Information and Accessibility**: Do you know where to seek guidance on the AI system and AI accountability requirements?
- **Operational Context (Design)**: Is the AI system used in the settings and circumstances for which it was designed?
- **Operational Context (Authorisation)**: Is the AI system being used within the settings and circumstances which have been authorised by your policing organisation?
- **Monitoring**: Do you have the means to monitor the performance of the AI system to check it remains in line with its intended purpose and quality requirements?
- **Intervention Protocols**: Do you know when and how to overrule AI decisions or recommendations, and when to escalate concerns regarding system behaviour?
- **Training Requirements**: Have you all the necessary information to carry out your task with the AI system?
- **Training**: Have you received sufficient training to be authorised to use this system and operate it lawfully?
- **Audit Trail**: Are your decisions for AI input/outputs or decisions based on AI inputs/outputs adequately recorded?
- **Mitigation Strategy**: Are appropriate measures in place to mitigate the impact of residual data or algorithmic bias?
- **Update and Modification Strategy**: Do you have a clear strategy for when and how to process updates or modifications?

*Oversight Bodies*

- **Compliance and Objective Alignment**: Does the AI system, its requirements and functionality achieve their stated purpose lawfully and effectively?
- **Ethical Governance Planning**: Are there any foreseeable ethical issues that may arise from using the AI system, and has a solution/mitigation roadmap been created?

- **Ongoing Assessment and Best Practice Guidance**: Are you monitoring and reviewing the AI system's lifecycle?

### *Theme 2: Data*

---

**Purpose**: This theme establishes an understanding of the data involved in creating the AI system (training/test data), as well as the data created for and by the AI system (inputs/outputs). It further ensures clarity about data management, risks and regulations, focusing on a number of critical considerations:

- Detailing the types and sources of data used in the development and deployment of the AI system, allowing proper evaluation.
- Establishing transparent procedures for lawful and appropriate acquisition, processing and management of data.
- Ensuring that legislative and policy compliance procedures in relation to data are effectively implemented and documented.
- Confirming the roles of responsible actors in relation to data are established, they are aware of their responsibility and possess appropriate skills and knowledge

**AI accountability principles addressed**: Lawfulness, proof, completeness, inclusivity, Explainability, transparency, enforceability and redress, impartiality, conduct, learning

---

## Data Guidance

### *Types and sources of data*

*Step 1: Detail the types and sources of data*

Understanding the source and nature of data used in the development of the AI system, as well as the data used by and produced by the AI system, is an important step in an informed assessment of the system and its outputs and impacts.

**To accomplish this, there should be a comprehensive record of**:

- The data used in the development of the AI system.
- Data used and generated by the AI system.
- Information on data relating to marginalised or vulnerable groups.

**Awareness of data biases**: 'As AI systems learn from data which may be unbalanced and/or reflect discrimination, they may produce outputs which have discriminatory effects on people based on their gender, race, age, health, religion, disability, sexual orientation or other characteristics' [6].

**To ensure accurate and consistent outputs it should be confirmed that**:

- Any bias present is known and documented.
- Appropriate mitigation measures against bias are in place where required.
- There is appropriate statistical weighting and validity for all data subjects.

These measures aim to establish a baseline of functionality for all individuals subject to the system.

*Step 2: Know the requirements for personal data*

The standard outlined by UK GDPR, including completion of a DPIA, remains relevant throughout the lifecycle of the AI system. It should be confirmed **whether the AI system processes sensitive or personal data at any stage of its lifecycle**.

**If so, the organisation should confirm**:

- The necessity and proportionality of all processing of personal data.
- The presence of procedures to inspect and control the processing of personal data.
- Systemic human intervention in the processing of personal data.
- Compliance of the AI system's processing of personal data with retention policies.

### *Data acquisition and management*

*Step 3: Set up clear data collection and processing protocols*

An important element for AI accountability is the **ability to justify the provenance and processing of data for and within the AI system.**

This includes data:

- Used in training, validation and testing of the AI system.
- Input data used by the AI system.
- Any data acquired from a third party or repurposed from existing datasets.

**In developing protocols and documentation about the acquisition and processing steps, it should be considered how the data was:**

| Chosen | Designed | Collected | Prepared* |
|---|---|---|---|

* Preparation may include annotations, labelling and cleaning of data

Each of the activities above represents a point of risk for the introduction of bias. An overview of how each process takes place allows confirmation of compliance with relevant standards both before and following the procurement of the AI system.

### *Data governance*

*Step 4: Assurance of relevant legislation and policy in the context of AI data*

**It should be confirmed that existing data governance processes within the organisation encompass the use of AI systems.**

Standard requirements include:

- Assurance of UK GDPR compliance.
- Completion of a DPIA, with regular updates.
- Appropriate Data Protection Policy.

These established practices should be assessed to confirm that internal policy is suitable for the use of the AI system. It may be the case that the existing policy is already appropriate. However, this evaluation and the grounds for it should be recorded. Where AI systems were developed before procurement, it should be confirmed that data protection compliance was built into the system design.

*Step 5: Ensure continued compliance with obligations for all data processors*

Written agreements should be signed to ensure all relevant data processors are aware of their commitments. This ensures that all parties are aware of their obligations. In the case of external parties, written contracts may also guard against the withdrawal of critical updates or data protection functions.

As with any IT system, building and maintaining compliance to adequate data practices is a core challenge. **In this respect, overarching knowledge and training requirements are**:

- Data protection awareness for all staff.
- Specific training for the AI system end-users/other accountable individuals.
- Integration within training of how to identify and report a personal data breach.

However, training alone is not sufficient. **To maintain compliance and ensure adaptation of data practices, there should be a regular review of**:

- Continued compliance with data protection policy and regulations.
- Effectiveness of data handling.
- Proper functioning of security controls, including cybersecurity.
- Effectiveness of the reporting and mitigation processes for data breaches.

### Data considerations throughout the AI lifecycle

*AI Accountability responsibilities for procurement*

- **Legality**: Ensure compliance with UK GDPR, the Equality Act 2010, ethical standards and other relevant requirements for data protection.
- **Data Sources and Compliance**: Conduct due diligence on those providing or developing the AI system and AI models (including data sources, bias mitigation, model transparency).
- **System Maintenance**: Include clear contractual obligations for continued system maintenance, where applicable (e.g. security, updates, audit access).
- **Risk Management**: Document intended use, risks and safeguards for all types of data, with special focus on personal and sensitive data.
- **Consultation**: Engage DPOs, legal and operational leads early; consult externally if data acquisition/processing is high-impact or otherwise a matter of public interest.

*AI Accountability responsibilities during deployment*

- **Oversight Responsibility**: Assign clear points of accountability for data oversight.
- **Access and Data Rights**: Set up clear responsibilities and rights for data accessing, processing, modification, deletion, etc.
- **Accuracy and Impact**: Monitor for bias, accuracy and disproportionate impacts of data practices, particularly for protected groups.
- **Data Protection**: Follow data handling, security and breach response laws and policies.
- **Security Protocols**: Establish clear protocols for handling data misuse and breaches from actions with or of the AI system, including reporting, mitigation and redress of the misuse/breach.
- **Learning**: Provide regular training and maintain audit-ready documentation on data practices and data management.

*AI Accountability responsibilities for migration/decommissioning*

- **Compliance**: Ensure that the discontinuation of the AI system complies with all existing data retention, deletion and data subject rights under UK GDPR.

- **Preservation**: Review whether any ongoing legal or public interest obligations require preservation of certain datasets or system outputs.
- **Data Storage**: Establish who has control to remaining/retained data, where it is located and how it can be accessed.
- **Archiving**: Archive necessary records for AI accountability and transparency, in a format proven to ensure longevity of data access.
- **Record Keeping**: Document reasons for retention or deletion of datasets.
- **Completeness**: Ensure third parties also remove stored data, where applicable.
- **Final Review**: Conduct a lessons-learned review of data practices to inform future AI procurements/use.

### Data considerations for intended users

*Chief Officers*

- **Strategic Accountability**: Are monitoring structures in place to ensure responsible data use and publication, in line with legal and ethical standards?
- **Lawfulness Assurance**: Are you confident that the planned processing of personal and sensitive data is lawful, especially when systems repurposed legacy data?
- **Governance Integration**: Do you have assurance that AI-specific requirements are incorporated into existing data governance structures, risk assessments and policies (e.g. UK GDPR, Equality Act 2010)?
- **Oversight of Development**: Are you confident that all data sources used in system development are known (including those from third parties)?
- **Bias and Fairness**: Do you have evidence that data bias has been adequately identified, assessed and mitigated against, particularly concerning vulnerable and marginalised groups?
- **Public Assurance**: Is the information about how the AI system uses data sufficiently available and suitable for public communications and community assurance?

*AI Users*

- **Data Protection**: Are you aware of how data protection requirements interact with the operation of the AI system?
- **Data Integrity**: Has all critical information been inputted into the AI system, allowing complete and honest outputs? **Awareness of Limitations**: Are you aware of any limitations or biases from training data and/or algorithms, which may affect the system?
- **Bias Vigilance**: Do data and accepted use policies ensure users remain alert to potential signs of erroneous or discriminatory outcomes, especially regarding protected characteristics?
- **Personal Data Handling**: Are procedures in place to ensure that all interactions with personal or sensitive data are lawful, necessary and proportionate?
- **Documentation Requirements**: Are you aware of requirements for record keeping in relation to data acquisition, processing, access and storage and grounds for decision-making?
- **Incident Reporting**: Do you know the correct method to report data protection breaches or unusual system outputs?
- **Training Compliance**: Have you received sufficient and required training on correct and safe data practices, including refresher training to remain certified?

*Oversight Bodies*

- **Audit Trails**: Do comprehensive records exist for all data used in the AI development and deployment, including third-party and legacy datasets?
- **Bias Detection Processes**: Are methods for identifying and mitigating bias in place, ensuring considerations for inclusion of minority/vulnerable groups?
- **Validation Evidence**: Is there sufficient statistical validity, weighting and test outcomes for AI models, particularly regarding fairness and generalisability?
- **UK GDPR and DPIA Oversight**: Is data protection compliance adequate and complete, particularly considering sensitive data use and system changes/evolution?

- **Compliance Monitoring**: Are regular internal reviews conducted on data handling, system updates and breach management?
- **Policy Adequacy**: Do internal policies sufficiently address AI-specific data risks and are they followed in practice?
- **Third-Party Accountability**: Has the policing organisation ensured external vendors comply with contractual obligations on data protection and system integrity?
- **Transparency and Publication**: Are there clear standards for documenting and, where appropriate, publishing information about AI data processes, consistent with national best practices?
- **Continuous Improvement**: Are procedures in place to ensure that improvements are an integral part of the review process, based on observed issues or evolving standards, including technological updates and procedural reforms?

### *Theme 3: Laws and Regulations*

---

**Purpose**: Lawfulness is an overarching principle in all areas of AI accountability. It ensures that the use of AI complies with applicable laws, codes of practice, regulatory requirements and policies. It further helps to clarify the legal context in which the system operates and safeguards public rights

**AI accountability principles addressed**: Lawfulness, proof, completeness, Explainability, enforceability and redress, conduct

---

## Laws and Regulations Guidance

### *Legal obligations*

*Step 1: Understand which laws apply to the AI system and current lifecycle stage*

A policing organisation must fully understand which laws, regulations and policies apply to the specific AI system, its purpose and application context.

**For this purpose, a policing organisation must:**

- Identify all applicable laws, regulations and policies for the designing/procurement, deployment and migration/decommissioning of the AI system, including any legal exemptions granted or safeguards enforced.
- Ensure the use of the AI system is necessary and proportionate for the given purpose (see example).
- Ensure relevant entries are recorded in the National Record of Processing Activities.

Different AI systems and application contexts will trigger different legal requirements. For example, an AI system managing a vehicle fleet will be subject to different regulatory frameworks than a system used for facial recognition.

Where there is no AI/purpose-specific law present, a policing organisation should still follow existing legal frameworks and align with national guidance and local policy.

When **working with international partners**, a policing organisation needs to be aware of, and possibly follow, other AI legal frameworks and regulations (such as the EU AI Act when working with agencies within Europe).

**Example—Breach of data protection laws [6]**

Under the Data Protection Act 1998, the Information Commissioner's Office (ICO) issued an Enforcement Notice to the Metropolitan Police Service (MPS) concerning its use of the Gang Violence Matrix (GVM), a tool designed to identify and assess the risk posed by individuals involved in gang-related violence across London. While the ICO acknowledged that the database served a legitimate purpose, it found that its application was unlawful and extended beyond what was necessary. As a result of the investigation, hundreds of individuals were removed from the matrix, and the GVM was eventually decommissioned by the MPS

*Step 2: Put procedures in place that monitor compliance with laws and regulations*

Once the applicable laws and regulations are identified, the policing organisation must take proactive steps to ensure compliance throughout the AI system's lifecycle. Each AI system may require a tailored compliance approach, depending on its purpose, impact and data used.

**To support monitoring of legal compliance policing should**:

- Ingrain laws and regulations that govern the AI system into monitoring and review procedures.
- Remain abreast of changes in the law and/or regulatory requirements.
- Conduct reviews of the AI system and its usage, when new laws and/ or regulatory requirements are enforced.

**Laws and Regulations considerations throughout the AI lifecycle**

*AI Accountability responsibilities for procurement*

- **Legal Compliance**: Document all applicable laws, regulations, policies and potential exemptions for the system's intended use.
- **Expert Review**: Ensure the AI system has been reviewed by a competent legal professional, to ensure compliance with relevant legislation.
- **Necessity and Proportionality**: Evidence that the AI use is both necessary and proportionate for its stated purpose.
- **Public Record**: Ensure the AI system is logged appropriately in the National Record of Processing Activities.
- **Alignment with Existing Standards**: Confirm alignment with internal and external governance frameworks (including national guidance, Codes of Practice and local policies).
- **Provider Disclosure**: Require external system providers to provide a full disclosure of data sources and system functionalities.
- **Provider Accountability**: Ensure third-party providers can demonstrate that the AI system has been developed in a lawful way and

does not contravene Codes of Conduct, ethics or any other policing guides and regulations.

- **Contractual Obligations**: Include legal compliance obligations and access to internal system documentation in provider contracts.

*AI Accountability Responsibilities during Deployment*

- **Designated System Owner**: The system sponsor is aware of their responsibility for system use, outputs, oversight and adherence to legal and ethical standards.
- **Legal Review**: Regularly review whether the system's usage still meets legal and policy requirements.
- **Reassessment Triggers**: Reassess the necessity and proportionality, if the system is repurposed or context changes.
- **Monitoring Procedures**: Implement compliance monitoring mechanisms to track and document legal adherence (e.g. audits, alerts for boundary violations).
- **Updating Public Record**: Ensure procedures are in place for updates to the National Record of Processing Activities, if data practices change.
- **Human Oversight**: Designate accountable personnel for legal oversight and establish a human-in-the-loop review process for high-impact decisions.
- **Incident Reporting**: Establish clear reporting lines for legal, policy or ethics breaches; maintain logs of legal, policy or ethics breaches, with a clear process for escalation, reporting and corrective action.
- **Remedy Procedure**: Establish responsibilities and mechanisms how to remedy legal and regulatory issues.

*AI Accountability responsibilities for migration/decommissioning*

- **Transition Process**: When decommissioning or transitioning systems, verify legal compliance will be maintained (e.g. data protection, fair processing).
- **Continued Compliance**: Ensure the AI system is no longer processing personal data unless justified and logged.

- **Preserve Documentation for Audit**: Archive legal assessments, data usage logs, compliance checks and decision records to support future reviews or legal inquiries in a format proven to facilitate long-term accessibility.
- **Data Handling**: Securely delete, anonymise or archive data in line with data protection regulations.
- **Full System Compliance**: Ensure third-party providers also comply with legal requirements during the migration/decommissioning processes, where applicable.
- **Final Review**: Conduct a final review of the AI system's performance in line with legislative requirements, and any lessons learned for future procurements/deployments.

## Laws and Regulations Considerations for Intended Users

### *Chief Officers*

- **Legal Status**: What is the current law relating to AI?
- **Applicable Frameworks**: Which laws, regulations, guidelines, policies and frameworks apply to the AI system and AI use case?
- **Cross-jurisdiction Use**: Will the AI system's outputs be provided to LEAs outside England, Wales, Scotland and Northern Ireland? If so, what other regulations and safeguards are to be followed?
- **Local Policy Validity**: What is the current organisational policy on using AI systems, and is it still in line with current law and regulations?
- **Policy Alignment**: Do organisational AI SOPs align with laws, regulations, guidance, policies and frameworks?

### *AI Users*

- **Permission Compliance**: Are all the necessary permissions to use the AI system in place?
- **Codes of Practice**: Are the correct Codes of Practice being followed using the AI system?
- **Lawful Use**: Is the AI system being used in a legal, proportionate and justifiable way?

- **Ethical Use**: Are you confident your use of the AI system is consistent with ethical policing standards, particularly regarding fairness, transparency and legality?

*Oversight Bodies*

- **Legal Compliance**: Are you able to assess, given the information available, that the correct laws and regulations are being followed and enforced?
- **Risk Communication**: Have you communicated legal risks and areas for improvement to the policing organisation?
- **Regulatory Review**: Given the information available, are you able to review processes that pertain to adherence to relevant laws and regulations for the AI system (e.g. is there an up-to-date DPIA)?

### *Theme 4: Risk Assessment and Management*

**Purpose**: This theme identifies risks, harms and biases associated with the AI system and implements mitigation measures. It covers both the impact of using and not using AI, with an emphasis on fairness and public safety. Identifying possible risk, harms and impacts of an AI system, as well as their likelihood is an important step to proactively manage these risks throughout the AI system lifecycle. It allows forward planning of how these risks (and associated harms) may be prevented, or where not possible, how they are to be managed and robust mitigation measures to be put in place

**AI accountability principles addressed**: Lawfulness, proof, completeness, Explainability, transparency, enforceability and redress, impartiality, conduct, learning

## Risk Assessment and Management Guidance

### *Risks and harms assessment*

*Step 1: Assess the risks and potential harms associated with the AI system and its use*

A comprehensive, proactive assessment of the potential risks associated with the AI system and its use helps the policing organisation to either prevent or prepare for their occurrence. The risk assessment can also establish whether deployment of an AI system poses a specific risk to particular communities and/or demographic groups.

**A risk register should be created, listing the type of risks, who they may affect (and associated potential harms), as well as their likelihood and potential severity:**

- **Risk Identification**: Identify risks associated with the AI system's procurement, deployment or migration/decommissioning (what could go wrong and to whom, e.g. individual, organisational, investigational integrity, community); also consider risks of not deploying the AI system.
- **Harm Identification**: Determine uses or outcomes of the AI system that could potentially cause harm; consider type of harms and who these harms may affect.
- **Likelihood**: Assess the likelihood of each identified risk; this assessment should be in line with methods used in wider organisational risk management.
- **Severity**: Assess the probable severity if the risk was realised; compare against established risk criteria (e.g. legislative requirements, organisational thresholds) to determine levels of significance, whether they are acceptable and whether they require mitigation.

The risk register should be reviewed on regular basis and after incidents.

---

**Example—Risks to specific communities [6]**
The Royal United Services Institute (RUSI) report 'data analytics and algorithmic bias in policing' [12] reviewed the use of data analytics and algorithms in policing within England and Wales. Their findings highlighted over-reliance on automation risked imbedding discrimination within policing practice.
The research highlighted concerns that algorithms could inadvertently perpetuate or amplify existing bias against particular groups, particularly if training data reflects historical bias in police practice, replicating historic assumptions. However, issues can emerge across the system lifecycle, including data collection, model development, deployment and evaluation

---

*Step 2: Conduct key assessments to review risks*

Conducting regular assessments about risks ensures that all foreseeable risks can be responsibly managed.

**Such assessments should consider whether measures**:

- Prevent or rectify data bias.
- Address security, privacy and confidentiality requirements.
- Identify and address unintended consequences of the AI system's data processing.
- Manage information risks.

**Several key assessments can help supporting this aim**:

- Data protection compliance assessments.
- Formal risk assessment.
- Ethics assessment (including compliance with the Code of Ethics).
- Equality Impact Assessment.
- Alignment with any established Codes of Practice.

A policing organisation should conduct specific **cyber security risk assessments,** to ensure the integrity of the AI system and any infrastructure it utilises.

*Step 3: Consider specific AI risks, misconduct and non-deployment*

**Particular AI risks**: There are a number of potential risks which may be of concern in police use of AI. While they are not specific to the technology, they may be exacerbated in this context.

| | | |
|---|---|---|
| Bias and discrimination | Risk: | AI systems trained on historical crime data can reflect and reinforce systemic biases, such as over-policing of certain communities |

(continued)

(continued)

| | | |
|---|---|---|
| | Example: | Predictive policing tools directing more proactive patrols to ethnic minority neighbourhoods, increasing arrests regardless of actual crime rates |
| False positives and inaccurate predictions | Risk: | Facial identification or predictive analytics can wrongly identify individuals as suspects, leading to detainment or surveillance |
| | Example: | Misidentification of an individual could result in the arrest of an innocent party |
| False negative and inaccurate predictions | Risk: | Facial identification or predictive analytics could fail to identify the offender as a suspect, leading to missed investigative opportunities |
| | Example: | A failure to identify an individual as a potential suspect (where the system reasonably would have been expected to do so) may lead to missed investigative opportunities, unnecessary use of further resources and loss of public confidence in the system |
| Lack of transparency | Risk: | Decisions made by opaque AI models may not be explainable to those affected or operating the system, reducing AI accountability |
| | Example: | A person may be denied bail from custody or labelled a flight risk without a clear reason that they or their legal representative can challenge |
| Automation bias | Risk: | Police officers or staff may over-rely on AI recommendations, treating them as infallible |
| | Example: | A predictive tool flags someone as a weapons threat, and officers presume the individual is dangerous without questioning the output |
| Function creep | Risk: | AI tools introduced for narrow purposes (e.g. identifying stolen cars) may be expanded over time to broader, more invasive uses |
| | Example: | A facial recognition system intended for serious crimes is later used to monitor the day-to-day movements of peaceful protestors |

(continued)

(continued)

| | | |
|---|---|---|
| Lack of accountability | Risk: | When AI goes wrong, it is often unclear who is responsible—The developer, the policing organisation or the individual officers |
| | Example: | Victims of algorithmic errors may struggle to seek redress due to complex responsibility chains |

**Misuse**: There are a number of potential risks posed by deliberate courses of action. Individual users may engage in misuse through unlawful access or processing of data. In this context, similar risk management strategies should be implemented.

A further consideration would be the capability of a user to deny an AI system critical information as a data input. Here outcomes could be manipulated by skewing the information reviewed by the AI system. Issues of this kind are likely to be system and context-specific but should be considered in the design process and usage policy.

**Non-use**: Where an AI capability is reasonably available for use, the implications of non-deployment must also be considered. This recognises that withholding AI in certain contexts may lead to negative outcomes or missed benefits. Considering the core policing duty of protecting life and limb, public safety requirements may dictate the use of the most effective lawful methods to achieve that aim.

Not engaging in AI use can thus represent a risk in and of itself, with missed opportunities to optimise policing activity, avert harm and provide a cost-effective service to the public. **Non-use should be an explicit part of the risk assessment and risk register**.

### *Risk and harm mitigation*

#### *Step 4: Define and regularly review mitigation measures*

**For each risk, mitigation measures should be included within the risk register.** Practical mitigation procedures will be system-specific, but may include:

- **Bias Auditing**: Testing of data and models for unfair outcomes or systemic bias.
- **Adversarial Testing**: Using red team testing and simulation to assess continued resilience.
- **Authentication and Authorisation**: Restrict who can access the system to those with a genuine purpose and appropriate training.
- **Responsible Usage Policy**: Define and enforce appropriate use, including clear procedures to prevent uncritical acceptance of AI-generated outputs.
- **Clear Responsibilities**: Every AI system in operation should have a clearly designated individual, accountable for the system's function, outputs and compliance with requirements of proper use.
- **Monitoring and Logging**: Utilise routine logging, dip sampling, system behaviour and usage patterns, to detect abuse or anomalies.
- **Human-in-the-loop**: Require human review for decisions impacting upon the public, i.e. a human must always serve as the ultimate decision-maker.
- **Fail-safes and Shutdown Procedures**: Implement mechanisms to interrupt or stop the system if it behaves unexpectedly.

*Step 5: Consider specific mitigation issues of personal data*

A particular matter of concern regarding the use of AI is its interaction with personal data. Deployers should be confident that:

- Any risk of bias in the processing of personal data has been mitigated.
- Other risks relating to the AI system processing personal data are addressed.

Special attention should be given to this topic with recorded assessments to detect and mitigate bias. Additional data protection and fairness risks should be reviewed continuously to ensure comprehensive safeguards are in place for all personal data processed by the system.

**Residual risks**: Where potential residual risks remain—despite legal compliance and procedures to address them—it should be evaluated whether such issues can be mitigated by procedure. Every reasonable effort should be made to eliminate all unlawful bias. However, no system is perfectly neutral, and it may be challenging to confirm total elimination in every use context. It is therefore important to include effective checks and safeguards within the method of deployment, in addition to data cleaning, to eliminate bias.

**Risk Assessment and Mitigation Considerations throughout the AI Lifecycle**

*AI Accountability responsibilities for procurement*

- **Clear Ownership**: Individuals or teams responsible for overseeing AI procurement decisions are given formal responsibility for managing the risks involved in deploying the AI system and the associated risk mitigation strategies.
- **Risk Forecasting**: AI system providers have communicated anticipated risks, including bias, privacy concerns and limitations of the system, and how they prevented, reduced or mitigated these risks in their system.
- **Ethics and Impact Evaluation**: Ethical reviews, equality reviews, equality impact assessments, etc., have been used to evaluate the suitability of the AI system for policing contexts generally, and the intended use case specifically.
- **Transparency Requirements**: Contracts are in place that include obligations for explainability, audit access and the provision of documentation on model functionality and data usage.
- **Beyond Value for Money**: Procurement decisions are based on demonstrable benefit to public safety and rights protection, not solely on cost-efficiency.
- **Comparative Risk**: Consideration should be given to risks associated with the specific system being considered compared to other available systems or no system being procured at all.

### *AI Accountability responsibilities during deployment*

- **Operational Oversight Framework**: Governance structures effectively oversee ongoing use, including human-in-the-loop decision-making and continuous evaluation, to monitor known and emergence of new risks.
- **Policy Implementation**: A responsible usage policy is in place, including user training, limits on automation and mandatory human review.
- **Bias and Impact Monitoring**: Methods are in place to monitor outputs for discriminatory patterns or systemic bias, particularly affecting vulnerable or historically disadvantaged communities.
- **Incident Reporting Mechanism**: Robust mechanisms are established for users, and where applicable the public, to report concerns or harms resulting from AI use.
- **Compliance Tracking**: The AI system effectively retains up-to-date logs of data use, decisions made and model performance for auditing and AI accountability purposes.

### *AI Accountability responsibilities for migration/decommissioning*

- **Residual Risk Review**: Lingering risks such as data retention, reputational damage or algorithmic bias in retained outputs have been identified and assessed.
- **Data Management**: The archiving process anonymises or deletes sensitive personal data in accordance with legal and policy requirements.
- **Lessons Learned**: Performance issues, stakeholder feedback and ethical challenges have been documented to inform risk management in future AI procurement and deployment.
- **Accountability Continuity**: Risk assessment and management measures remain active through the decommissioning and post-use assessment phases.
- **Transition Planning**: When replacing the AI system, risk mitigation, ethical review and user retraining are included in the transition plan.

## Risk Assessment and Mitigation Considerations for Intended Users

*Chief Officers*

- **Strategic Accountability**: Have you ensured clear assignment of responsibilities, with a named individual accountable for risk assessment, monitoring and mitigation?
- **Governance Framework**: Are you confident that robust governance structures are in place that support the effective risk assessment and management?
- **Deployment Risk Assessment**: Have appropriate pre-deployment risk assessments been conducted (e.g. ethics, data protection, equality impact assessment, etc.)?
- **Bias Mitigation**: Are reasonable measures in place to eliminate or mitigate bias across the AI lifecycle?
- **Policy Enforcement**: Are measures, such as the Responsible Usage Policy, and requirements for mandatory human oversight and decision-making, practically enforceable?
- **Operational Decision-Making**: Does your organisation have the tools to evaluate both the risks of using and not using AI?
- **Resource Allocation**: Are there sufficient resources for continuous risk monitoring, auditing and evaluation?
- **Community Impact Considerations**: Are ongoing measures in place to assure mitigation of disproportionate impacts on particular communities or demographic groups?

*AI Users*

- **Understand Limitations**: Do you avoid uncritical acceptance of AI outputs and decisions?
- **Human Oversight Role**: Do you recognise that users are responsible for final decisions and should not defer accountability to the AI system?
- **Bias Awareness**: Are you aware of potential biases in outputs and the steps needed to question and validate findings?

*Oversight Bodies*

- **Audit and Evaluation**: Have you incorporated the use of AI expressly within your internal and external audit arrangements?
- **Bias and Fairness**: Does your organisation have staff with sufficient technical understanding to scrutinise the AI system, identify bias and assess mitigation measures? (This may include assessment of data, algorithms and outcomes.)
- **Documentation and Transparency**: Do you have clear requirements for policing organisations on the documentation of usage policies, decision-making and AI accountability structures?
- **Ethical Compliance**: Do existing standards for ethical behaviour effectively encompassed police use of AI?
- **Misconduct and Misuse**: Do oversight and redress processes have effective methods of monitoring and identifying signs of system misuse (e.g. unlawful access, intentional data skewing) and ability to assess whether safeguards are effective?
- **Non-deployment**: Do oversight processes account for whether a decision not to deploy AI in appropriate contexts represents a risk to public safety or service quality?
- **Risk Assessments**: Can you verify that formal risk assessments, ethics reviews and data protection compliance measures account for AI use?

### *Theme 5: Oversight and Redress*

**Purpose**: This theme identifies risks, harms and biases associated with the AI system and implements mitigation measures. It covers both the impact of using and not using AI, with an emphasis on fairness and public safety. Identifying possible risk, harms and impacts of an AI system, as well as their likelihood is an important step to proactively manage these risks throughout the AI system lifecycle. It allows forward planning of how these risks (and associated harms) may be prevented, or where not possible, how they are to be managed, and robust mitigation measures to be put in place

**AI accountability principles addressed**: Lawfulness, proof, Explainability, transparency, compellability, enforceability and redress, impartiality, constructiveness, conduct, learning

## Oversight and Redress Guidance

### *Oversight process*

*Step 1: Identify oversight requirements and process*

Ensuring adequate oversight throughout the AI lifecycle is an integral part of AI accountability. Oversight refers to actions taken to review and monitor AI policies, plans and implementations, with the aim to ensure that (1) they achieve expected results, (2) represent value for money and (3) are compliant with relevant policies, laws, regulations and ethical standards.

**A first step is to map out oversight requirements and related oversight procedures, i.e.**

- Legal requirements for oversight.
- The areas and aspects that require oversight.
- The procedures and mechanisms to implement and sustain adequate oversight.
- The internal and external functions and organisations responsible for oversight (see Step 2).
- The information required to allow adequate oversight (see Step 3).

*Step 2: Understand the type and nature of information being shared*

**A policing organisation needs to identify the oversight bodies that will be responsible for monitoring, assessing and enforcing the appropriate use of the AI system in their particular context, including potential conflicts of interests (see Theme 6 for considerations of stakeholder interests).**

Oversight functions can be internal or external to the police organisation, as well as with a formal (legally mandated) or non-formal role.

Examples of internal groups:

- Functions ultimately responsible for the success of the AI efforts.
- Individuals and policing functions using the AI capabilities Examples of external groups:
- Legally mandated oversight bodies.
- Groups most affected by the AI system and its deployment.
- Communities concerned with the issue the AI deployment addresses.
- Press and media organisations.

For the implementation of an effective oversight process, internal and external oversight functions need to work in alignment.

**To accomplish this, a policing organisation should look to**:

- Clarify the role and authority of the oversight bodies.
- Understand the expected level of transparency and reporting.
- Determine if there are any conflicts of interest.

External bodies will vary depending on context. A non-exhaustive list of entities which may be involved in AI accountability oversight is provided in Sect. 7 'governance bodies with relevance for AI accountability'.

*Step 3: Understand the type of information to be shared*

The oversight process will require sharing of disparate types of information with each of the oversight bodies and groups.

**A non-exhaustive list of information that may typically be requested for oversight purposes is**:

- Purpose and scope of the AI system.
- Data.
- Model design and training.
- Deployment context.

- Decisions and outputs.
- User behaviour.
- Post-deployment monitoring.
- Governance processes.
- Risk register.
- Impact assessments.

In practicality, answering these questions will require creation and retention of materials such as:

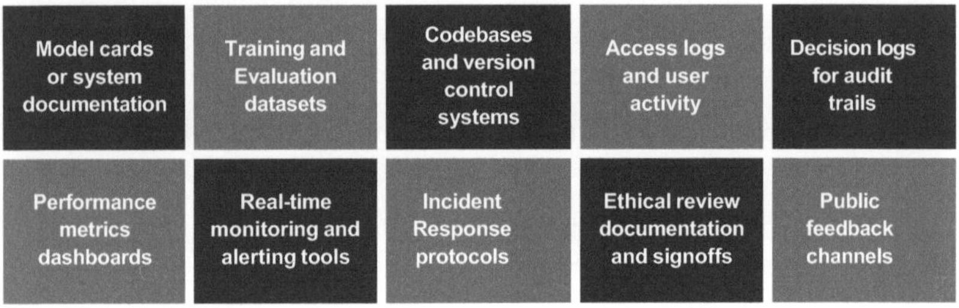


Some information may be subject to a legal requirement to share, other information may be in the public interest to provide (e.g. for transparency purposes to retain public trust). Rationales, requirements and potential limitations to the sharing with each oversight body and group should be internally transparent. This transparency is also linked to Risk Management (discussed in Theme 4).

**Sharing restrictions**: Knowing what information a policing organisation can share will influence which and how information can be provided to different oversight bodies. If operational and security requirements restrict the ability to share information, police organisations still need to ensure that the AI system and its procurement, use, etc., will undergo scrutiny by appropriate independent assessors with appropriate clearance.

**To help process this requirement a policing organisation should know**:

- The operational or security requirements restricting the ability to share information to oversight bodies.

- If it is in the public interest to share information, and if so, what information should be shared (e.g. algorithmic information, datasets, limitations of the AI system, outputs, deployment details).

*Step 4: Identify how you are sharing the information*

Once a policing organisation knows what information to share and who to share it with, communication mechanisms need to be established. These need to be targeted towards the respective audience (e.g. internal vs external oversight functions).

AI systems can be extremely complex and may require a data scientist or specialists to explain technical aspects. A policing organisation needs to ensure that the information is provided in a way the intended recipient will be able to understand; or there is a risk that the information is interpreted incorrectly and causes ill-informed or incorrect judgement on the AI system.

**A policing organisation should further ensure that there is a clear method to feedback and record recommendations and requests made by an oversight body**.

***Redress process***

*Step 5: Establish a redress process*

Redress ensures that there are consequences and/or remedies in case things go wrong. This includes appropriate corrections within the police organisation, AI system, its usage, etc., but can also extend to compensation or other forms of remedy. Adequate redress procedures provide fairness and integrity and are a cornerstone of AI accountability.

**An effective redress process includes**:

- Information that affected individuals can understand, regardless of their technical or legal literacy.
- Availability in multiple languages and formats, depending on audiences (e.g. online forms, via call centre staff, in-person support).

- Transparency regarding what decisions were made or influenced by an AI system.
- Explanations of how the decision was reached and the role the AI system played.
- Escalation procedures.
- Procedures that operate promptly, with clear timelines for response and resolution.
- Impartial review, ideally by decision-makers who are independent of the original decision.
- Clear lines of responsibility for final outcomes, including implementation of corrections or compensation.
- Maintenance of an auditable record of actions taken.

*Step 6: Monitor the redress process*

**The redress process will need to be reviewed and revised** to ensure it remains fit-for-purpose, efficient, compliant and aligns with any change to legal or operational requirements. Lessons identified in the redress process should also be communicated to training and development teams, if applicable.

**Erroneous outcomes produced by the AI system must be escalated to the developers/third-party for correction through a formal escalation process.** This approach ensures the policing organisation maintains a central understanding on the scale and scope of any issues and can implement contingency measures if necessary.

It is important to focus on the purpose of accountability processes, to prevent a counterproductive and unresponsive form of compliance seeking. Regular and transparent evaluation ensures alignment with emerging risks and standards, ultimately fostering public trust in the process.

### Oversight and Redress Considerations Throughout the AI Lifecycle

*AI Accountability responsibilities for procurement*

- **Oversight Role Identification**: Identify relevant oversight bodies and their role (ICO, ethics boards, HMICFRS, etc.).
- **Information Sharing Limits**: Determine operational/security limitations on information sharing.

- **Information Definition**: Define what AI-related information can be shared (e.g. algorithms, datasets, protocols).
- **Redress by Design**: Include redress mechanisms in the AI system design.
- **Lead Assignment**: Assign responsibility for oversight and redress to internal leads.

*AI Accountability responsibilities during deployment*

- **Engagement Channels**: Establish targeted engagement channels with oversight bodies and core stakeholders.
- **Public Transparency**: Where possible, update public-facing information to help build trust.
- **Redress Monitoring**: Establish and monitor the redress process and adjust where necessary.

*AI Accountability responsibilities for migration/decommissioning*

- **Oversight Notification**: Notify all relevant oversight bodies before migration or decommissioning.
- **Redress Resolution**: Resolve outstanding redress issues.
- **Data Assurance**: Confirm all data has been securely removed or migrated.
- **Final Review**: Conduct final oversight review of the AI system's performance.
- **Documentation Archiving**: Archive any relevant documentation such as decision logs and redress records.
- **Asset Register Update**: Update the information asset register accordingly.

## Oversight and Redress Considerations for Intended Users

*Chief Officers*

- **Oversight Identification**: Have the correct and relevant oversight bodies been identified for this AI system and use purpose?

- **Redress Mechanism**: Is there an accessible and appropriate redress process in place that specifically accounts for harms or disputes arising from the AI system?
- **AI Accountability Leads**: Are there designated responsible leads within the organisation for managing oversight and redress processes?
- **Feedback Integration**: Are findings, recommendations or requirements (such as lessons learnt or directives) from oversight bodies systematically fed back into operational practice and policy?

*AI Users*

- **Information Restrictions**: Are operational and security-based restrictions on information sharing clearly documented, and do you understand how they apply to your role?
- **Error Identification**: Can you identify and report anomalies, biases or errors in the AI system—including those caused by user behaviour or input?
- **Reporting Compliance**: Are you following reporting procedures (internal and external) when an issue or concern related to AI deployment arises?
- **Operational Feedback**: Are you actively providing operational feedback to support oversight reviews and improve system performance?
- **Disclosure Protocol**: Are you complying with organisational policy on information disclosure, especially with regard to oversight bodies and sensitive or classified information?

*Oversight Bodies*

- **System Review**: Is the review of the AI system's use and outputs fully independent and unfettered?
- **Compliance Monitoring**: Does the policing organisation demonstrate compliance with relevant laws, regulations and policies?
- **Transparent Communication**: Does communication of findings regarding the AI system fit into existing feedback structures?
- **Oversight Integrity**: Are the independence, expertise and impartiality of oversight activities assured by sufficient ability to interpret the technical aspects of AI use?

- **Redress Effectiveness**: Are redress mechanisms in place and evaluated to ensure their continued relevance and effectiveness?

### *Theme 6: Identification and Management of Stakeholders*

---

**Purpose**: This theme aims at mapping and engaging all relevant parties who are involved in or affected by the AI system. This includes those internal to the policing organisation, external parties such as AI developers or oversight bodies and members of the public. A thorough identification and management of stakeholders ensures that:

- A full picture of which relevant groups, bodies and organisations are responsible for AI accountabilities (e.g. oversight bodies, affected communities, AI providers) and their respective roles is available.
- A proactive management strategy for how to communicate and engage with these stakeholders in effective ways is in place along with a clear strategy for how to respond to challenges, information requests or audit

**AI accountability principles addressed**: Lawfulness, completeness, inclusivity, Explainability, transparency, compellability, enforceability and redress, impartiality, constructiveness, conduct

---

## Identification and Management of Stakeholder Guidance

### *Identification of AI accountability stakeholders*

*Step 1: Identify all stakeholders relevant for this AI lifecycle phase*

Relevant stakeholders are those that either:

- Use the AI capability or outputs of the AI capability for policing purposes.
- Have responsibility to make design, procurement, deployment and/or decommissioning decisions.
- Are performing oversight functions.
- Are affected by the AI capabilities.
- Need to be informed about the AI procurement, deployment, etc., for other reasons.

**For a comprehensive identification consider the 4 categories** (see Sect. 7 for a non-exhaustive list of examples):

| | | | |
|---|---|---|---|
| Police-internal functions | Governance bodies | Specific communities and general public | External organisations and expert groups |

**Guiding questions that a policing organisation should consider include:**

- What criteria are used to identify relevant stakeholders?
- Do any of the stakeholders have protected characteristics that need special treatment/consideration?

*Step 2: Ensure that stakeholders are reviewed at important points of the AI lifecycle and if needed updated*

Different stakeholders will become relevant at different points of an AI lifecycle. A policing organisation can optimise efficiency by ensuring that they are engaging with the correct stakeholders and to the correct extent.

**To achieve this, a policing organisation should consider**:

- Which points of the AI lifecycle are relevant to which stakeholders?
- What communication is required and to what extent?
- What risks arise if the stakeholder is not properly engaged with?

***Management of stakeholders***

*Step 3: Define the communication and engagement requirements for each stakeholder, how this will be done and who is responsible for the communication/engagement*

A policing organisation needs to consider their audience when communicating to stakeholders and ensure that the communication is received as intended. Active stakeholder management enables clear communication, coordinated decision-making and early identification of concerns. Although a policing organisation will decide what it communicates and when, transparency is encouraged during all engagement to help build trust, clarity and understanding.

**Communication and engagement requirements can be**:

- To inform or require information from a stakeholder.
- To provide or request evidence, materials or documentation.
- To support financial or legal accountability.

**Guiding questions to consider are**:

- Are there access restrictions across stakeholder groups for the sharing of information, decisions, data, etc.?
- Can they be explained/evidenced?

*Step 4: Account for stakeholder groups that will require additional considerations to communicate or engage with and how to ensure that they still receive adequate information*

Diverse populations will receive and understand information differently; therefore, it is crucial that the policing organisation takes measures to reduce potential barriers or challenges.

**Example challenges can be**:

- Low trust towards police.
- Special requirements for communication or access to information, including language or impairments/disabilities and available technology.

*Step 5: Define how successful communication and engagement will be assessed and put countermeasures in place, in case they fall short*

To establish meaningful stakeholder communication and engagement, it is helpful to create criteria that communications and engagement efforts can be evaluated against. This will help to ensure that messages are clear, impactful and appropriate for the intended audience.

**A policing organisation should consider whether**:

- Communication has successfully increased awareness.
- The communication cycle includes information on how concerns have been addressed appropriately.
- The communication strategy addressed negative impacts and information about complaints/redress procedures.

The measurement of effect needs to be quantifiable and link the exposure (of the message) to an outcome (e.g. sentiment towards the police and the AI system). Typically, this can be achieved through community engagement.

*Step 6: Outline potential concerns and expectations of general public/society that may cause challenges to the AI deployment and how to address them*

Public concerns such as privacy invasion, bias and intrusive surveillance can significantly influence the acceptance or rejection of an AI system. By identifying these challenges early, developers and policymakers can take proactive measures to address them through design, transparency and engagement.

**A policing organisation should seek to**:

- Identify all possible concerns that the public may have with the police use of the AI system.
- Address concerns before they become an issue (control the narrative).
- Have contingency communication plans in place, particularly if the engagement reveals a high level of opposition.

## Stakeholder considerations throughout the AI lifecycle

*AI Accountability responsibilities for procurement*

- **Stakeholder Identification**: Identify relevant stakeholders and include them in the procurement process (e.g. oversight bodies, police procurement officers, legal advisors, external topic experts).
- **Communications Planning**: Create a communication plan in order to engage with relevant stakeholders, including the public about the intended AI system use.
- **Guidelines**: Articulate the intended aims and scope of the AI system use into a formal project proposal.
- **Due Diligence**: Ensure that all parties involved in the development and provision of the AI system have been identified and appropriately assessed.
- **Stakeholder Involvement**: Ensure that legal, technical, ethics and community oversight stakeholders have been engaged early in procurement deliberations.

*AI Accountability responsibilities during deployment*

- **Stakeholder Communication Protocol**: Identify if any stakeholders should be informed of deployments, on what timescales and by what means.
- **Operational Insight and Data Sharing Requirements**: Establish if any stakeholders require insight into decisions or access to ongoing operational data.
- **Final Deployment Responsibility**: Identify who is accountable for final deployment decisions, including pausing or stopping deployments.
- **Data Retention and Documentation**: Identify what information, data or materials need to be recorded or retained, under what authority and in what format, to ensure long-term AI accountability towards relevant stakeholder groups.
- **Public Trust**: Address identified public concerns about police use of AI through transparency and communications strategy.
- **Transparency and Disclosure**: Ensure a process for affected stakeholders to be informed, if an adverse incident takes place.

- **Feedback and Complaint**: Ensure affected stakeholders can inform police about (potential or actual) negative impacts.

*AI Accountability responsibilities for migration/decommissioning*

- **Stakeholder Continuity**: Identify stakeholders relevant to decommissioning decisions, including the impact of withdrawal or replacement of the AI system.
- **Communications Planning**: Identify who needs to be informed about migrations and decommissioning, including their implications and impact.
- **Ongoing Accessibility**: Identify who will have, or needs, access to data, information and materials from decommissioned systems and likely time scales.
- **Data Protection**: Maintain processes to facilitate deletion of data as required by data protection legislation, policy or in response to requests from affected stakeholders.
- **Public and Internal Communication**: Clearly communicate the system's decommissioning to stakeholders and affected individuals, especially where trust, rights or services are involved.

## Stakeholder considerations for intended users

*Chief Officers*

- **Stakeholder Mapping**: Are you confident that a comprehensive and up-to-date mapping of all relevant stakeholders across the AI lifecycle has taken place?
- **Internal Collaboration**: Have you mandated cross-departmental reviews to capture diverse internal and external perspectives in stakeholder identification?
- **Responsibility Assignment**: Have clear responsibilities been assigned within the Chief Officer team, and to relevant leaders in the organisation, for managing stakeholder communication and engagement?
- **Inclusive Engagement**: Are strategies in place to effectively engage communities with protected characteristics or low trust in police?

- **Public Expectations**: Have you identified potential concerns and expectations from the public and accounted for this in both AI deployment and corporate communication?
- **Evaluation Measures**: Are there measures for assessing communication and engagement success, and are there procedures for corrective action, when standards are not met?

*AI Users*

- **Stakeholder Awareness**: Are you aware of who the relevant stakeholders are, and how your use of the AI system impacts them?
- **Access Restrictions**: Do you understand and follow the access restrictions related to information sharing across different stakeholder groups?
- **Information Flow**: Are you clear on your role in passing information to oversight bodies or responding to public or governance queries?
- **Community Interaction**: Are you prepared to engage appropriately with affected communities, recording, reporting and actioning any concerns they express about AI use?
- **Process Updates**: Are you involved in or informed of updates to stakeholder management processes that could affect your operational use of the AI system?

*Oversight Bodies*

- **Stakeholder Identification Review**: Have you verified that the police have conducted an inclusive and systematic identification of all relevant stakeholders?
- **Lifecycle Triggers**: Are mechanisms in place to ensure stakeholder reviews are triggered at key lifecycle stages (e.g. moving from piloting to deployment)?
- **Transparency Check**: Have you assessed whether communication and engagement strategies are transparent and fair across all stakeholder groups?

- **Equality and Diversity**: Have you evaluated how the policing organisation ensures effective engagement with all of the communities it serves?
- **Documentation Audit**: Is there clear, accessible documentation showing how stakeholder voices have been integrated into decision-making and governance?
- **Public Concern Oversight**: Are you monitoring how the police identify and respond to societal and public concerns regarding AI deployment?
- **Public Engagement**: Are appropriate community input and transparency used to build trust in AI use in policing and prevent alienation or suspicion?

### *Theme 7: Accountability Evidence*

**Purpose**: This theme ensures that evidence, about how AI accountability is assessed and captured, is robust, verifiable, accessible and understandable. This is not about criminal evidence but how policing evidences its AI accountability. By knowing how to prove AI accountability correctly, a police organisation will be better positioned to respond to challenges. The process will also highlight areas of AI accountability, which may need developing further

**AI accountability principles addressed**: Lawfulness, proof, completeness, inclusivity, Explainability, transparency, compellability, enforceability and redress, constructiveness, conduct, learning

## Accountability Evidence Guidance

### *Creation of robust evidence*

*Step 1: Define the criteria for robust and sufficient evidence*

Although there might be criteria that are specific to each policing organisation, or even each AI system and its intended use, **the criteria that can guide what should be considered as AI accountability evidence should align with**:

- Requirements by law.
- Additional regulatory requirements.
- The Code of Ethics requirements.

- National (NPCC) Strategy, Policy and Guidance.
- Regional/Local Policy and procedure.
- Criminal evidence integrity.
- Requirements of stakeholders and oversight bodies.

A policing organisation will need to decide which information best demonstrates AI accountability to capture and document the relevant and appropriate evidence ('proof') across all themes.

**This evidence of AI accountability, or proof, will be important in numerous contexts**. For example, it can be used to support operational cases through criminal justice processes demonstrating, and where required, the legality and integrity of the AI system in question. Proof will also assist in maintaining public confidence and consent, through the transparent demonstration of AI systems being used, their function and their utility.

**Example—failing to acquire proof [8]**

In bridges v South Wales police, the court of appeal ruled that South Wales police failed to comply with the public sector equality duty when deploying facial recognition technology as they did not take reasonable steps to investigate whether the system could produce biased outcomes based on race or sex. Although there was no clear evidence that the system was biased, the failure to assess for possible biases was itself a breach of duty.

*Step 2: Build processes to collect AI accountability evidence*

To ensure AI accountability evidence is being recorded, a policing organisation should seek to embed AI accountability collection methods into the AI systems lifecycle, including the AI system itself. Standardised procedures that are consistent across departments should be developed and should be co-created with support from governance, legal and technical teams to ensure the right data is being recorded and to maximise usability.

**When creating a process to record AI accountability, a policing organisation should consider including data such as:**

- **What** AI accountability evidence needs to be recorded?
- **When** does it need to be recoded, and for how long?
- **Where** is the data going (has it been sanitised; can it be disclosed)?
- **Why** is the data recorded (does it add value to service accountability)?
- **How** is the data recorded (is it accessible, usable, secure and archived correctly)?

At each stage of the AI system's lifecycle, it is recommended that AI accountability proof is reviewed to ensure completeness.

### Safeguarding evidence

*Step 3: Enable correct accessibility*

A policing organisation must ensure that the evidence used to demonstrate AI accountability for their AI system complies with any legal or policy requirements that govern information/material handling and disclosure. However, a balance needs to be struck in that the proof of accountability must also be accessible to all relevant stakeholders and oversight bodies in order to support transparency and effective oversight. The information communicated needs to be in a clear and understandable format, ensuring that both technical and non-technical audiences can use the information appropriately (see Theme 6).

To ensure evidence of AI accountability is both useful and accessible, a police organisation should follow the MTbD principle in how that information is communicated and shared. This means tailoring information such as algorithmic documentation, deployment records or risk assessments to the needs and technical understanding of various oversight bodies and stakeholders.

**A controlled access framework should be established so that the right stakeholders can view the necessary information without compromising operational sensitivity or data protection obligations.**

## AI Accountability Evidence considerations throughout the AI life-cycle

*AI Accountability responsibilities for procurement*

- **Procurement Decision Records**: All decisions related to the procurement of the AI system must be formally recorded; this should incorporate necessity considerations.
- **Third-Party Disclosure Requirements**: Developers or third-party providers must meet disclosure requirements, including providing technical information such as training data.
- **Demonstrating Accountability Evidence**: The policing organisation must be able to demonstrate to stakeholders and oversight bodies that it is collecting and maintaining the correct AI accountability evidence.
- **Procurement Disclosure Safeguards**: Control measures must be established to protect against unnecessary or unlawful disclosure of procurement-related information.

*AI Accountability Responsibilities during Deployment*

- **Deployment Evidence Recording**: The organisation must record appropriate AI accountability evidence throughout the deployment of the AI system.
- **Ongoing Transparency**: Standards of transparency must be maintained to demonstrate proper evidence collection to stakeholders and oversight bodies.
- **User Interaction Monitoring**: User interactions with the AI system must be monitored and auditable to ensure responsible use and system traceability.
- **System Update Logging**: All updates to the AI system must be logged and justified.
- **Operational Disclosure Safeguards**: Measures must be in place to prevent unnecessary or unlawful disclosure during system operation.
- **Records of Performance Evidence**: Evidence related to system performance, audit trails and decision logs must be retained in line with local policy and legislative requirements.

*AI Accountability Responsibilities for Migration/Decommissioning*

- **Migration and Decommissioning Records**: The organisation must record and review all steps involved in the secure migration or decommissioning of the AI system.
- **Archiving Performance Evidence**: Evidence related to system performance, audit trails and decision logs must be archived appropriately.
- **Accessibility of Archived Data**: Archived data must be accessible in case decisions need to be revisited.
- **Documentation of Decisions**: All decisions made during migration/decommissioning must be documented and justified.
- **Disclosure Safeguards**: Safeguards must remain in place to prevent unlawful or unnecessary disclosure or retention during and after migration/decommissioning.

**AI Accountability Evidence considerations for intended users**

*Chief Officers*

- **Assigned Responsibilities**: Have clear responsibilities been assigned for recording and maintaining accountability evidence at all stages of the AI system's lifecycle?
- **Oversight for External Scrutiny**: Is there oversight in place to ensure decisions made under your authority can be explained and justified?
- **Periodic Review Mechanisms**: Have mechanisms been put in place to periodically review AI accountability evidence?

*AI Users*

- **Awareness of Evidence Requirements**: Are you aware of what types of AI accountability evidence you are expected to (and not to) record?
- **Access to Systems**: Do you know how to access the tools and systems required to document your use of the AI system?

- **Reporting Procedures**: Are procedures in place for you to record unexpected behaviours, impacts or performance issues in the AI system or user error?
- **Data Migration and Decommissioning**: Have the steps for secure data migration or decommissioning been recorded and reviewed?
- **Disclosure Controls**: Are there control measures in place to protect against unnecessary or unlawful disclosure?

*Oversight Bodies*

- **Access to Records**: Do you have access to comprehensive, understandable records showing how and why the AI system is procured/used/decommissioned?
- **Documentation Provided**: Has the policing organisation provided you with sufficient technical and operational documentation to enable meaningful scrutiny?
- **Information Format Suitability**: Is the format in which information is shared suited to your review and reporting requirements?
- **Evidential Integrity**: Are you able to verify the integrity of AI accountability evidence, and trace how decisions about it were made and recorded?
- **Alignment with Standards**: Can you assess whether the AI system's use aligns with legal, ethical and policy standards, and whether gaps in AI accountability have been identified and addressed?

### *Theme 8: Adaptability and Learning*

**Purpose**: Adaptability and learning ensures that policing stays up-to-date with technological innovations or requirements in AI and societal changes. It encourages continuous improvement through feedback, review and learning. It thus ensures that:

- Procedures and skills are reviewed and updated where necessary.
- Those developing or using the AI system remain aware of their responsibilities.
- Lessons are learned from mistakes, inefficiencies or unexpected/negative consequences

**AI accountability principles addressed**: Lawfulness, inclusivity, transparency, compellability, enforceability and redress, conduct, learning

## Adaptability and Learning Guidance

### *Continuous improvement*

*Step 1: Embed learning and adaptability early*

As governance requirements, AI technologies and social attitudes change, a policing organisation will need to keep current with their demands and needs. For an AI system to remain current and safe, continuous improvement needs to be designed into all parts of the AI lifecycle and ecosystem. Early adoption of the requirement for learning and adaptability will help a policing organisation become dynamic and responsive to changes outside of their control. **Adaptability and learning thus need to be designed in throughout the AI lifecycle**. Contractual agreements to third parties, for example, should explicitly make provisions for the need of an AI system to be responsive and unplanned updates may be required.

**Consideration should be given to**:

- Where along the AI lifecycle adaptability will be required.
- Development of a roadmap for lessons to be identified across the AI lifecycle to support learning (individual or organisational), e.g. on:
  - Procurement contracts.
  - Tradecraft.
  - The AI system.
  - Processes.
  - Management and governance.

*Step 2: Up-skill the organisation*

A general AI awareness will be required of staff. This is similar to generalised data protection or fire safety knowledge requirements expected of the full workforce and should provide awareness of general principles of responsibility for using AI accountably. This prepares policing for a future where AI use is deeply imbedded into routine organisational functions and operational policing. It also ensures policing has an understanding of how external parties using AI may impact upon them. This awareness should include knowledge of monitoring processes and contextualised understanding of the consequences for breaches.

**To achieve this, a policing organisation should consider:**

- Training for users (including managers).
- Raising awareness for those not directly involved.

*Step 3: Continually review, feedback and adapt*

Processes should be put in place, where a policing organisation can record lessons learnt that can then be fed back into the AI ecosystem. The process will need to include lessons identified during all stages of the AI lifecycle.

**To support this process, the organisation should:**

- Ensure there is a learning feedback loop at all stages of the AI lifecycle.
- Assign responsibility to manage and implement changes (e.g. does this information need to be provided to training and development teams or to AI developers?)
- Monitor and adjust to new changes to ensure they are effective and correct.

## Adaptability and Learning throughout the AI lifecycle

*AI Accountability responsibilities for procurement*

- **Training Information**: Confirm providers/developers provide sufficient system information for an effective training course on the system.
- **Future Proofing**: Embed requirements for ongoing system adaptability as a core procurement requirement.
- **Ongoing Improvements**: Guarantee long-term developer involvement and lifecycle support.
- **Adaptability**: Include adaptability metrics in system evaluation, such as capacity for retraining, patching, as well as integration of feedback.

*AI Accountability responsibilities during deployment*

- **Training Assurance**: Implement structured training, communicating requirements and expectations appropriate to the user's role.
- **Base Level Knowledge**: Maintain a core level of organisational AI knowledge delivery provision to all staff, as well as specific training for AI system users.
- **Integration of Feedback**: Confirm feedback channels for reporting issues, errors or suggested updates to result in meaningful improvement.
- **Proactive Monitoring**: Ensure monitoring processes proactively identify areas in which the system is functioning sub-optimally, allowing adaptation and mitigation.
- **Collaborative Improvement**: Maintain continuous collaboration between developers, deployers and end users to integrate updates and learning.
- **Continual Development**: Avoid treating deployment as a static endpoint, and build in mechanisms for ongoing improvement and responsiveness to emerging risks.

*AI Accountability responsibilities for migration/decommissioning*

- **Organisational Memory**: Ensure any new or replacement AI system benefits from cumulative learning, rather than a functionality reset with each new AI system.
- **Reflective Practice**: Use migration or decommissioning as an opportunity to reflect on the effectiveness of feedback and improvement mechanisms, to benefit future systems.
- **Full Lifecycle Evaluation**: Evaluate how the system adapted (or failed to adapt) over time, as part of knowledge transfer and organisational learning.
- **Comprehensive Review**: Included all involved parties, including developers, users and other stakeholders in the review process.

## Adaptability and Learning considerations for intended users

*Chief Officers*

- **Building Base Level Knowledge**: Is AI awareness built into routine organisation-wide training (akin to data protection or fire safety)? Do the users of AI systems within the organisation have sufficient knowledge to fulfil their roles and accountability requirements appropriately and comprehensively?
- **Setting Expectations**: Have you effectively and consistently communicated that adaptability and continuous learning are part of expected AI use within your organisation?
- **Support Feedback Structures**: Are you confident there are robust mechanisms for user feedback and that issues raised are taken seriously and escalated where needed?
- **Sustain Developer Engagement**: Are agreements in place for your enabling services teams to have long-term collaboration with developers, to maintain and improve the system post-deployment?
- **Anticipate Change**: Does your review process ensure adaptability in reaction to external developments (e.g. emerging vulnerabilities of the AI system, regulatory shifts, etc.) in both risk management and operational strategies?
- **Proactive Change**: Do regular system reviews proactively identify methods of improved practice, both technical and procedural?

*AI Users*

- **General Awareness**: Have you developed a baseline understanding of AI and sufficient knowledge of the AI systems you will directly use, including ethical, legal and AI accountability considerations?
- **Ongoing Training**: Have you participated in required training, including refresher training, to understand evolving system capabilities and expectations?
- **Feedback Responsibility**: Do you know how to report issues, errors or misuse through clearly defined channels, whether system-integrated or organisational?

- **AI Accountability in Practice**: Do you understand the specific consequences of misuse or process breaches for the AI system you use?
- **Adaptive Use**: As the AI system evolves, are you able to adjust practices to maintain effectiveness and integrity?
- **Understand Monitoring**: Are you aware of how system use is monitored and how this relates to performance and compliance evaluations?
- **Documented Processes**: Do you know how to access process documentation?

*Oversight Bodies*

- **Training Content**: Does the syllabus and its ethos meet required standards of quality assurance?
- **Equality**: Are considerations relating to diversity and equality in AI use effectively covered by training?
- **Continuous Review**: Do organisational accountability processes exist to ensure the police organisation remains responsive to technological and legislative developments?
- **Evaluate Feedback Process**: Are feedback mechanisms effective and accessible, resulting in demonstrable improvements?
- **Culture of Compliance**: Is there meaningful engagement with AI accountability requirements, fostering trust and transparency?
- **Improvement Transparency**: Does the organisation maintain documentation and records evidencing open communication about system performance and updates to training or feedback processes?
- **Threat Benchmarking**: Does the organisation respond to emerging threats and assess the system's resilience against them?

## Core Accountability Stakeholders

Please note: The overview is not intended to be exhaustive and will differ depending on the context of use and UK Legal Jurisdiction(s) the system is used within.

**Police-internal AI Accountability stakeholders**

- Chief Constable/Chief Officers.
- Enabling Services, such as.
  - Procurement Professionals.
  - IT Department.
  - Information Management.
  - Data Assurance/Protection.
  - (AI) Product Sponsor
  - Corporate Communications.
  - Legal Services.
  - Training Professionals.
  - Users deploying the AI system.
  - Functions that make decisions based on AI outputs.
  - Functions that provide data for AI training or AI deployment.
  - Staff otherwise impacted by the AI system.

**Governance bodies with relevance for AI Accountability [5]**

*England and Wales*:

- His Majesty's Inspectorate of Constabulary and Fire and Rescue Services (HMRCFRS).
- Independent Officer for Police Conduct (IOPC).
- HM Courts and Tribunal Service.
- Local Elected Policing Bodies.

*Northern Ireland:*

- NIPB.
- His Majesty's Inspectorate of Constabulary and Fire and Rescue Services (HMRCFRS).
- Criminal Justice Inspection Northern Ireland.
- Police Ombudsman's Office.
- HM Courts and Tribunal Service.
- Chief Human Rights Commissioner.
- Local Elected Policing Bodies.

*Scotland*

- SPA.
- His Majesty's Inspectorate of Constabulary in Scotland.
- Police Investigations and Review Commissioner.
- Scottish Information Commissioner.
- Local Elected Policing Bodies.

*Cross-Jurisdictional*

- Information Commissioner (England and Wales, Northern Ireland, limited involvement in Scotland).
- Investigatory Powers Commissioner (IPC).
- Equality and Human Right Commission.

*External organisations and bodies:*

- Industry/developers from which AI capabilities are procured.
- Interest groups, such as:
  - Business groups, Chambers of Commerce, Trade Unions.
  - Private security.
  - Transport providers.
- Equalities groups.
- Press and media.

**Public and special communities**:

- Communities with special interests, such as.
- Children and youth.
- Groups with disabilities.
- Rural communities.
- Religious communities.
- Environmental groups.
- Specifically impacted or affected communities, including minorities and vulnerable groups.
- Groups with Protected Characteristics (Equality Act 2010).

## Glossary

**Advanced Data Analytics** The use of complex techniques and tools, such as machine learning, predictive modelling or statistical analysis, to extract insights, identify patterns, or support decision-making from large or complex datasets.
**AI (Artificial Intelligence)** The capability of a machine to perform tasks that typically require human intelligence, such as learning, reasoning or decision-making.
**AI System** A system that uses AI technologies, such as machine learning, to perform tasks, make decisions or generate outputs, often with a degree of autonomy.
**AI Ecosystem** The interconnected network of all entities (e.g. people, data, processes, legal and regulatory frameworks) that are influenced by, or influence the AI lifecycle, including outputs.
**AI User** An individual or department that deploys, operates or relies on the outputs of an AI system. This includes those involved in initial procurement, local oversight, implementation or direct use of the system in decision-making.
**Algorithm** A set of rules or instructions given to an AI system to help it learn, solve problems or perform tasks.
**Bias (in AI)** A systematic error in an AI system that can lead to unfair outcomes, often due to flawed data or assumptions.
**Machine Learning (ML)** A subset of AI where systems learn from data and improve their performance over time without being explicitly programmed.

## References


1. B. Akhgar, P.S. Bayerl, 'How to do AI accountability – and why it's worth the effort'. In Higgins, A., Halkon, R. (eds.). Policing and the Fourth Industrial Revolution: Cumberland Lodge Police Conference 2024 – Conference Report. London: The Police Foundation, pp. 16–17 (2024). Available at: https://www.police-foundation.org.uk/wpcontent/uploads/2010/10/cumberland-lodge-2024.pdf
2. B. Akhgar, P.S. Bayerl, G. Mounier, R. Linden, B. Waites, AP4AI: Accountability principles for artificial intelligence in the internal security domain. Eur. Law Enforcement Res. Bull., Special conference edition **6**

(2022). Available at: https://shura.shu.ac.uk/31123/. Accessed 13 Mar 2025

3. College of Policing, *Code of Ethics* (2024). Available at: https://www.college.police.uk/ethics/code-of-ethics. Accessed 11 May 2025
4. House of Lords Library, *Policing in the UK: Governance, Oversight and Complaints (Online)* (House of Lords Library, London, 2020) Available from: https://lordslibrary.parliament.uk/research-briefings/lln-2020-0013/. Accessed 26 Mar 2025
5. Information Commissioner's Office (ICO), *What about Fairness, Bias and Discrimination?* (2023). Available at: https://ico.org.uk/for-organisations/uk-gdpr-guidance-and-resources/artificial-intelligence/guidance-on-ai-and-data-protection/how-do-we-ensure-fairness-in-ai/what-about-fairness-bias-and-discrimination/. Accessed 28 Apr 2025
6. Information Commissioner's Office (ICO), *Metropolitan Police Gangs Matrix* (n.d.). Available at: https://ico.org.uk/for-the-public/ico-40/metropolitan-police-gangs-matrix/. Retrieved 29 Apr 2025
7. Judiciary of England and Wales, *R (Bridges) v Chief Constable of South Wales Police and others: Press Summary* (Online) (Judiciary UK, 2020). Available at https://www.judiciary.uk/wp-content/uploads/2020/08/R-Bridges-v-CC-South-Wales-ors-Press-Summary.pdf. Accessed 30 Apr 2025
8. National Police Chiefs' Council (NPCC), *Covenant for Using Artificial Intelligence (AI) in Policing* (2023). Available at: https://science.police.uk/site/assets/files/4682/ai_principles_1_1_1.pdf. Accessed 18 Apr 2025
9. C. Novelli, M. Taddeo, L. Floridi, *Accountability in artificial intelligence: what it is and how it works*. AI &Society, **39**, pp.1–12 (2023). https://doi.org/10.1007/s00146-023-01635-y
10. Office of the Police Chief Scientific Adviser (OPCSA), *UK Police Industry Charter* (Online) (2024). Available at: https://science.police.uk/site/assets/files/5011/police_industry_charter.pdf. Accessed 11 May 2025
11. Office for Artificial Intelligence, *Guidelines for AI Procurement* (Online) (HM Government, 2020). Available at: https://www.gov.uk/government/publications/guidelines-for-ai procurement/guidelines-for-ai-procurement. Accessed 30 Apr 2025
12. Royal United Services Institute, *Data Analytics and Algorithmic Bias in Policing*. London: RUSI. (2019). Available at: https://assets.publishing.service.gov.uk/media/5d7f6b2540f0b61ccdfa4b80/RUSI_Report_-_Algorithms_and_Bias_in_Policing.pdf

# Index

## A





B. Akhgar and P. S. Bayerl (eds.), *Accountability for Artificial Intelligence in Policing and Security*, Security Informatics and Law Enforcement, https://doi.org/10.1007/978-3-032-41284-3

**T**



**U**

MIX
Papier aus verantwortungsvollen Quellen
Paper from responsible sources
FSC® C105338

If you have any concerns about our products,
you can contact us on
**ProductSafety@springernature.com**

In case Publisher is established outside the EU,
the EU authorized representative is:
**Springer Nature Customer Service Center GmbH**
**Europaplatz 3, 69115 Heidelberg, Germany**

Printed by Libri Plureos GmbH
in Hamburg, Germany